FM 3-75 (FM 3-05.50)

Ranger Operations

May 2012

Headquarters, Department of the Army

Field Manual	Headquarters
No. 3-75 (FM 3-05.50)	Department of the Army
	Washington, DC, 23 May 2012

Ranger Operations

Contents

Figures

Tables

Preface

Field Manual (FM) 3-75, *Ranger Operations*, establishes doctrine for Army Special Operations Forces (ARSOF) Ranger operations. It describes Ranger roles, missions, capabilities, organization, mission control, employment, and sustainment operations across the range of military operations. This manual is a continuation of the doctrine established in the Joint Publication (JP) 3-05 series and FM 3-05, *Army Special Operations Forces*.

PURPOSE

FM 3-75 describes the Ranger strategic landscape, fundamentals, core tasks, capabilities, and sustainment involved in the full range of military operations. This manual serves as the doctrinal foundation for subordinate Ranger doctrine, force integration, materiel acquisition, professional education, and individual and unit training.

SCOPE

This manual describes the principles, fundamentals, guidelines, and conceptual framework to facilitate interoperability and the doctrinal foundation for the development of subsequent tactics, techniques, and procedures; doctrine; and training literature. This manual complements and is consistent with joint and Army doctrine. The focus of this manual is on the operational level of Ranger operations. Although FM 3-75 focuses primarily on the Ranger Regiment and battalion, it also addresses ARSOF units from the Ranger platoon to the United States Army Special Operations Command (USASOC).

APPLICABILITY

Commanders and trainers should utilize this manual, as well as other related publications, in conjunction with theater mission letters, command guidance, and unit mission-essential task lists (METLs) to plan and conduct mission-specific training. The key to ARSOF mission success is to plan and practice operations before executing an assigned mission. This publication applies to the Active Army, Army National Guard (ARNG)/Army National Guard of the United States (ARNGUS), and United States Army Reserve (USAR) unless otherwise stated.

ADMINISTRATIVE INFORMATION

This manual contains numerous acronyms, abbreviations, and terms. Users should refer to the Glossary at the back of this manual for their meanings and definitions. Unless this publication states otherwise, masculine nouns and pronouns do not refer exclusively to men. The proponent of this manual is the United States Army John F. Kennedy Special Warfare Center and School (USAJFKSWCS). Submit comments and recommended changes to Commander, USAJFKSWCS, ATTN: AOJK-CDI-CIC-JA, 3004 Ardennes Street, Stop A, Fort Bragg, NC 28310-9610, or by e-mail to JAComments@soc.mil.

Chapter 1

Introduction

The Ranger Regiment is a direct reporting unit of USASOC. Within this publication, the term "Ranger force" refers to any size force consisting of units from the 75th Ranger Regiment. Ranger force operations support the United States (U.S.) national policies and strategic objectives by providing the President, the Secretary of Defense, and the geographic combatant commanders with a ground force capable of conducting joint special operations direct action throughout a range of military operations. The Ranger force is organized as an airborne light infantry regiment capable of rapidly deploying and operating anywhere in the world, in any weather or terrain. This highly trained special operations force infiltrates by land, sea, or air, and functions effectively in complex joint and interagency environments. The Ranger force successfully conducts missions requiring the precise application of combat power against targets across the operational spectrum. In addition to being trained to function under a wide range of conditions, the Ranger force specifically prepares to operate in politically sensitive environments, under restrictive rules of engagement, and usually at night. To ensure success, the all-volunteer Ranger force is manned through a rigorous selection and assessment process. This results in an exceptionally talented and motivated force of Soldiers led by exceptionally qualified and experienced leaders.

TASK

1-1. The 75th Ranger Regiment plans and conducts special military operations against strategic and operational targets in pursuit of national or theater objectives. Rangers conduct military operations independently, as well as jointly with conventional forces, other special operations forces (SOF), or interagency forces. Rangers perform complex infantry missions that conventional infantry, airborne, or air assault units may lack the specific expertise to perform. Ranger direct-action operations are usually deep-penetration raids or interdiction operations against targets of strategic or operational significance. Ranger direct-action operations may be conducted during low-intensity conflicts within an operational environment of conventional or coalition units and with close coordination between the units and the Rangers. Direct-action missions for Rangers typically include forced-entry operations, airfield seizures, and the capture or destruction of targets. Rangers conduct operations in accordance with the commander's intent and are not constrained to or prohibited from any types of missions.

1-2. Rangers are rapidly deployable airborne light infantry organized and trained to conduct highly complex operations either with or in support of other special operations units of all Services. Additionally, Rangers may execute direct-action operations in support of conventional missions conducted by a combatant commander. Rangers can operate as conventional light infantry when properly augmented with other elements of combined arms.

1-3. The Ranger Regiment is the ARSOF light infantry force specially organized, trained, and equipped with the capability to deploy a credible military force quickly to any region of the world. Rangers perform specific missions with other SOF, and often form habitual relationships. These missions differ from the missions of conventional infantry forces in the degree of risk and the requirement for precise, discriminate use of force. The Ranger Regiment uses specialized equipment, operational techniques, and modes of infiltration and employment.

ORGANIZATION

1-4.   The headquarters of the Ranger Regiment (Figure 1-1) is similar to the headquarters of other Army brigade combat teams. In addition to commanding and controlling three Ranger infantry battalions and the Ranger Special Troops Battalion, the regiment headquarters may, if augmented, exercise operational control of conventional forces, logistical assets, and other SOF. The Ranger Regiment is supported by a 2-man Air Force Special Operations Command (AFSOC) special operations weather team (SOWT), and has an established relationship with a USAF air support operations squadron for joint terminal attack controller support.

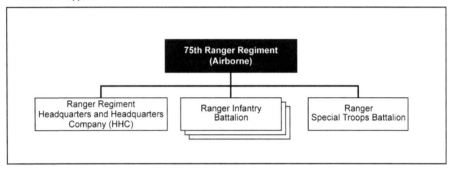

Figure 1-1. 75th Ranger Regiment

1-5.   The Ranger infantry battalion is the primary combat element within the regiment. It is structured similar to an airborne light infantry battalion (including mortar, reconnaissance, and sniper platoons), but without an antitank company. The Ranger infantry battalion consists of an HHC, four Ranger infantry companies, and a Ranger support company. The battalion has an attached USAF tactical air control party and each rifle company has three rifle platoons as well as a headquarters section.

1-6.   The Ranger Regiment provides a liaison team with secure communications to the supported commander's headquarters as needed. The liaison team provides the coordination of operations and logistics at the supported headquarters. An intelligence liaison officer (LNO) is placed at the theater joint intelligence center or the supported unit's all-source intelligence center to ensure the exchange of accurate and timely intelligence between the regiment and supported units.

1-7.   Because Rangers do not have the organic capability to establish their own operations bases, they normally exercise command and control through command posts collocated on bases with other SOF or conventional units. The Ranger Special Troops Battalion provides the unique logistics, communications, reconnaissance, and command and control support required to ensure success.

EMPLOYMENT CONSIDERATIONS

1-8.   For rapid deployments in support of contingency operations, Ranger units are self-sustaining for up to 72 hours. Beyond 72 hours, Ranger units need extensive, dedicated, and responsive external support throughout all phases of training and operations. The Ranger Special Troops Battalion and Ranger support companies at the battalion level provide the regiment internal logistical support, thus allowing the regiment to sustain combat operations beyond 72 hours.

1-9.   Because Rangers emphasize highly complex offensive operations, the commanders should not assign missions that conventional light infantry units can perform.

1-10.   The Ranger Regiment has a reconnaissance company within the Special Troops Battalion, as well as a reconnaissance platoon in each of the Ranger battalions. The reconnaissance company can conduct a full range of missions from tactical to strategic reconnaissance and surveillance.

1-11. Rangers—unlike Special Forces (SF), Military Information Support (MIS), and Civil Affairs (CA)—are globally oriented, rather than regionally oriented. Current force structure and contingency requirements preclude their apportionment to a specific geographic combatant commander. Rangers can deploy worldwide when their military presence would better serve U.S. interests.

1-12. Rangers may participate in joint or multinational training exercises with allied or friendly military forces. This participation enhances U.S. national interests by demonstrating the capabilities of ARSOF.

STRUCTURE, MANNING, AND EQUIPMENT

1-13. The Ranger force is roughly structured along the lines of a brigade combat team with a robust regiment headquarters and four battalions. The Ranger Special Troops Battalion includes reconnaissance, signal, operations, and military intelligence companies. The regiment has three subordinate Ranger infantry battalions, each consisting of four rifle companies, a support company (Echo Company), and an HHC. Each battalion HHC contains an organic reconnaissance platoon, a sniper platoon, and a mortar platoon. The Ranger force is organized, equipped, and trained to fight at the Ranger section, platoon, company, battalion, or regiment level, but it has the flexibility to provide tailored elements to joint special operations task forces (JSOTFs) or other headquarters. Additionally, the Ranger force has the flexibility to employ other conventional forces or SOF placed under Ranger mission command.

1-14. The Regimental Assessment and Selection Program—which screens, assesses, selects, and indoctrinates Rangers into the force—is the key to forming the solid foundation and vast capabilities of the Ranger force. The Ranger Operations Company in the Ranger Special Troops Battalion is the proponent for the Ranger Assessment and Selection Program. Officers, noncommissioned officers (NCOs), and Soldiers of all military occupational specialties are actively recruited from across the Army to join the Ranger ranks. New recruits are physically and psychologically assessed prior to being assigned to a Ranger unit. Recruited officers have successfully served as platoon leaders, company commanders, special staff officers, or battalion commanders in conventional units before their initial assignment to the Ranger Regiment. Enlisted Rangers reporting from initial-entry training or another unit are assessed similarly and indoctrinated before being assigned to the regiment headquarters or to a Ranger battalion.

1-15. The Ranger force undergoes continuous, intense training to ensure the capability of being resourced for training and the ability to be sheltered from outside distracters. The training focuses on attaining and maintaining extremely high levels of proficiency on core skills to produce exceptionally capable individual Rangers and small Ranger units by leveraging the experience of the Ranger leaders. The higher-level training emphasizes conducting sophisticated operations in joint environments.

1-16. The Ranger force is largely dependent upon outside sources for strategic or operational mobility and fire support. However, the Ranger force is specially equipped with weapons, night vision, communications, and other equipment which allow it to leverage state-of-the-art technology. Additionally, the Ranger tactical ground movement capabilities include the ground mobility vehicle-Ranger (a SOF-version of the up-armored high-mobility multipurpose wheeled vehicle), the Stryker infantry carrier vehicle, and other armored vehicles. It is the combination of sophisticated tactics, techniques, and procedures with these additional capabilities that provide the force with the enhanced mobility, lethality, and survivability.

1-17. The Ranger force has the capability to—

- Plan and conduct joint special operations in conjunction with Army, Air Force, Marine, and Navy SOF.
- Conduct or support a forcible entry in conjunction with other joint special operations or conventional assets to establish lodgment for inserting follow-on forces deep in enemy or denied territory by airborne assault or helicopter air assault.
- Conduct sustained combat operations in support of the combatant commanders.
- Provide liaison teams for up to four higher controlling headquarters. (Each team is equipped and staffed to communicate with each command's deployed Ranger unit and to integrate the Ranger units into the warfighting functions of the supported command.)
- Employ cross-functional teams to serve as intermediate headquarters between company-size elements and battalion headquarters. (These teams are task-organized in accordance with

mission, enemy, terrain and weather, troops and support available, time available, civil considerations [METT-TC] to perform numerous functions—for example, fusing intelligence and operations, performing liaison, conducting command and control, synchronizing with conventional forces, and executing the complete targeting cycle.)

- Maintain a regiment command and control element, a reconnaissance element, and a battalion in an alert posture prepared for deployment within 18 hours of notification—Ranger Ready Force 1. (One company from the same battalion maintains an initial-ready company on a 9-hour notice-to-move posture.)
- Employ light, medium, or heavy mortars in accordance with METT-TC.
- Employ sniper teams in support of tactical operations to increase operational area security and to minimize collateral damage with precision fires during limited visibility.
- Employ reconnaissance teams at regiment or battalion level to find enemy forces and to shape direct-action raids and other missions or to conduct special reconnaissance in support of Ranger operations.
- Conduct synchronized combination infiltration by fixed- or rotary-wing aircraft, tilt-rotor aircraft, ground-mobility platforms, and dismounted to perform precision raids and sensitive site exploitation.
- Conduct urban combat. (Rangers are highly trained in urban combat and operate primarily at night, maximizing the advantages of state-of-the-art technology for night vision and target acquisition. Rangers operate under very restrictive rules of engagement to minimize collateral damage and noncombatant casualties.)
- Compress the military decision-making process and troop-leading procedures.
- Create an environment in which other SOF have freedom to operate.
- Conduct operations to safeguard and evacuate U.S. citizens or to protect U.S. property abroad.
- Provide a Ranger deployable planning team on short notice to any warfighting commander, joint task force (JTF), or JSOTF headquarters to plan potential Ranger operations in support of an emerging or ongoing contingency operation.
- Conduct terminal guidance operations against high-value targets, either in support of operations by a larger Ranger force or as the primary Ranger mission in support of direct-action operations conducted by other forces.
- Move small Ranger elements, small numbers of evacuees, supplies, or casualties through urban terrain in armor-protected vehicles.
- Operate exclusively in a digital environment maximizing situational awareness and understanding of every Ranger.
- Operate in a chemical, biological, radiological, and nuclear (CBRN)-contaminated environment for up to 72 hours in conjunction with other JSOTFs.
- Conduct CBRN sampling to detect and identify CBRN contamination.
- Conduct limited decontamination for up to the 14 personnel within the Emergency Personnel Decontamination System; with logistical support, conduct operational decontamination—mission-oriented protective posture gear exchange.
- Provide control for equipment and personnel decontamination sites on airfields or other targets.
- Embark naval vessels with special operations rotary-wing aircraft to launch air assault operations to strike littoral targets.
- Provide a quick-reaction force capable of executing time-sensitive missions using multiple insertion techniques.

Chapter 2
Organization

The Ranger Regiment is structured similarly to a brigade combat team. The Ranger Regiment contains a dynamic regimental headquarters, a Ranger Special Troops Battalion (which includes reconnaissance, communications, operations, and military intelligence companies), and three Ranger battalions.

RANGER REGIMENT

2-1. The 75th Regiment headquarters, 3d Battalion, and the Ranger Special Troops Battalion are located at Fort Benning, Georgia. The 1st Battalion is located at Hunter Army Airfield in Savannah, Georgia, and the 2d Battalion is located at Fort Lewis, Washington. Figure 2-1 depicts Ranger unit locations.

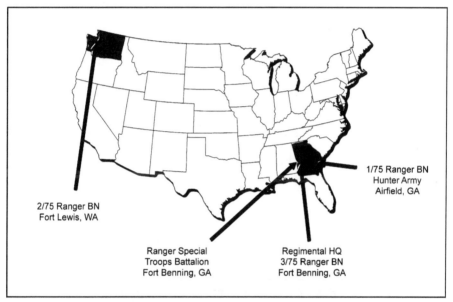

Figure 2-1. 75th Ranger Regiment locations

2-2. The Ranger Regiment headquarters plans, coordinates, synchronizes, integrates, and exercises mission command of operations for its subordinate battalions. The staff of the HHC is organized in accordance with current doctrine located in Appendixes C and D of ADP 6-0, *Mission Command*. The regimental staff (Figure 2-2, page 2-2) has the personnel and equipment for three standing command and control packages:

- *Special Operations Task Force (SOTF)*. The SOTF package is a robust mission command and control node that focuses on supporting large-scale regimental deployments and operations. This package provides the required structure for the regiment to mission command and control a SOTF and provides augmentation to joint headquarters, as required.

- *Assault mission* command and control. This package, commonly referred to as the STRIKE package within the regiment, is a specially tailored command and control node. It is manned by military personnel and focuses exclusively on supporting national-level mission command and control requirements.

- *Rear tactical operations center (home-station operations center [HSOC])*. The rear tactical operations center, also known as the HSOC, ensures the regiment maintains a sustained garrison mission command and control capability to train, field, and equip personnel. The rear tactical operations center ensures continuity of effort during regimental surge operations and frees military personnel for tactical employment.

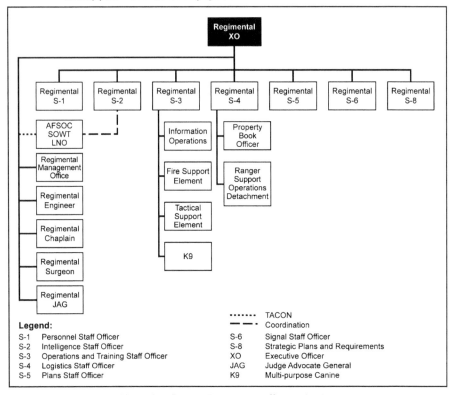

Figure 2-2. Ranger Regiment staff organization

RANGER SPECIAL TROOPS BATTALION

2-3. The Ranger Special Troops Battalion comprises four companies: the reconnaissance, military intelligence, communications, and Ranger Operations Company. The battalion is supported by a command and staff structure that is responsible for providing administrative and training oversight to these organizations. The Ranger Special Troops Battalion commander performs the rear tactical operations center commander duties when the regimental staff is deployed. The Ranger Special Troops Battalion staff is organized as depicted in Figure 2-3, page 2-3.

2-4. The Regimental Reconnaissance Company comprises a headquarters element and an operations, intelligence, signal, support, and selection and training section. The company has a total of six reconnaissance teams. The seven-man teams are made up of individuals selected from among the regiment's most mature and experienced NCOs. These Soldiers endure a rigorous selection and training process before being assigned to the Regimental Reconnaissance Company.

2-5.   The Regimental Military Intelligence Company includes intelligence personnel commanded by a military intelligence captain. The company comprises an all-source analysis platoon, single-source section, counterintelligence/human intelligence (CI/HUMINT) section, and an intelligence, surveillance, and reconnaissance section. The company has the capability to task organize in support of the subordinate Ranger battalions and the regimental command and control nodes.

2-6.   The Regimental Communications Company comprises command post support, maintenance, nodal support, and network support sections that are commanded by a signal captain. Rangers use and develop cutting-edge single-channel and multichannel communications.

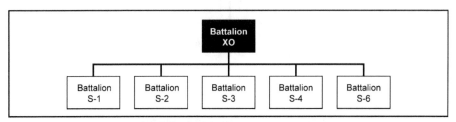

Figure 2-3. Ranger Special Troops Battalion staff organization

2-7.   The Ranger Operations Company manages three phases of Ranger assessment and training as follows:

- *Ranger Assessment and Selection Program 1.* Junior enlisted Rangers (private through sergeant) who have been newly assigned to the regiment attend this 8-week assessment. They learn the regiment's standing operating procedures and are required to meet the mental, physical, and morale requirements before being assigned to a battalion.
- *Ranger Assessment and Selection Program 2.* Newly assigned NCOs (staff sergeant through sergeant major) and officers attend this phase of training.
- *Small Unit Ranger Tactics.* Selected Rangers undergo this 3-week phase of training for final preparation to succeed in Ranger School.

RANGER BATTALION

2-8.   There are three identical Ranger battalions subordinate to the Ranger Regiment. Figure 2-4, page 2-4, illustrates the organization of each battalion.

2-9.   The battalion HHC staff is organized in accordance with the current doctrine located in Appendixes C and D of ADP 6-0. The battalion HHC comprises a medical section, a communications section, a reconnaissance platoon, a mortar platoon, and a sniper platoon.

2-10.   The reconnaissance platoon is organized into a headquarters section and four reconnaissance teams. These six-man reconnaissance teams are selected from the most capable and experienced Rangers in the battalion. Reconnaissance teams have a habitual relationship with the four rifle companies.

2-11.   The battalion mortar platoon can operate by split section or as a platoon and is organized into a headquarters section, two fire-direction center sections, and two sections comprising two mortar squads each. The mortar platoon has the 60-millimeter (mm), 81-mm, and 120-mm mortar systems. Mortars are issued and employed based on METT–TC using the same concept as the arms room.

2-12.   The battalion sniper platoon is organized into a headquarters section and four sections comprising three two-man sniper teams. The sniper teams employ both antipersonnel and antimateriel sniper systems with advanced night-vision capability. Sniper teams have a habitual relationship with the four rifle companies.

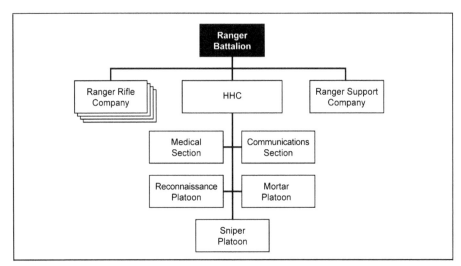

Figure 2-4. Ranger battalion organization

RANGER RIFLE COMPANY

2-13. Each Ranger battalion has four rifle companies. The rifle companies are organized and equipped parallel to the battalion structure. These rifle companies are capable of employing limited special operations, as well as conventional ground-mobility assets with various mounted weapons systems. Having four distinct rifle companies enables the battalion to task-organize four cross-functional teams. Figure 2-5 shows the rifle company organization.

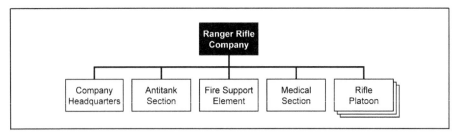

Figure 2-5. Ranger rifle company organization

2-14. The antitank section is organized into three distinct two-man antitank teams. The antitank section employs the 84-mm Carl Gustav Ranger Anti-Armor Weapon System and the Javelin. The fire support element consists of an artillery forward observer and a radio-telephone operator for each platoon. The company headquarters has an artillery fire support officer, a fire support NCO, and a fire support specialist. The medical section consists of one medical NCO and one enlisted medic for each platoon, as well as two medical NCOs assigned to the company headquarters.

2-15. Each rifle platoon consists of three rifle squads and a weapons squad. The nine-man rifle squad has two MK46 light machine guns and two M320 40-mm grenade launchers. The weapons squad has three MK48 machine gun teams consisting of three Soldiers to a team. Rifle platoons are capable of operating in two sections, allowing the Ranger force commander the flexibility to mass forces. Each section is supported with communications specialists, forward observers, medical personnel, and heavy weapons organic to the platoon.

RANGER BATTALION SUPPORT COMPANY (ECHO COMPANY)

2-16. Echo Company is the support company located at each Ranger battalion. Echo Company provides direct logistics support and sustainment of Ranger operations. It logistically support all of the unit's training, and provides logistics and force health protection (FHP) to operations conducted by other forces, as directed. Echo Company includes a headquarters with a distribution, sustainment, and maintenance platoon; and a property book section.

2-17. Echo Company can rapidly organize and deploy all of its assets to provide logistics support to a battalion and the regimental headquarters. These multifunctional support companies are organic to each of the three battalions within the Ranger Regiment and provide field maintenance as follows:

- Class I, II, III (P) (B), IV, V, VII, VIII, and IX supplies.
- Water production and limited distribution.
- Transportation.
- Aerial delivery.
- Bare-base support.
- Property management.
- Food service.

RANGER EQUIPMENT FORCE MODERNIZATION PROGRAM

2-18. The 75th Ranger Regiment has an aggressive force-modernization program that ensures Ranger forces are equipped with the most technologically advanced weapons and equipment possible. The program is led by the regimental strategic plans and requirements section (S-8), using USASOC's Planning, Programming, and Budget Execution System to ensure funds are allocated to procure new systems for future required Ranger force capabilities. Force-modernization program personnel work directly with the USASOC Combat Development Manpower and Programming section, G-8, and also coordinate with the U.S. Army Infantry Center staff responsible for testing and fielding new equipment. The Ranger force-modernization staff often coordinates with elements of the U.S. Army Infantry Center Futures Group, including the Directorate for Combat Developments, U.S. Army Training and Doctrine Command Systems Manager—Soldier, the battle lab, and others.

2-19. The regiment's force modernization task is to identify and address required Ranger capabilities to enhance mission performance. The program focuses on the shoot, move, communicate, and survive systems. Ideas for new equipment, enhancements, and improved interoperability also result from routine training and operations between Rangers and other SOF.

2-20. The intent of the force modernization program is to—

- Improve the operational capability of the Ranger Regiment by increasing lethality, mobility, survivability, and situational understanding of the individual Ranger through the modernization of equipment.
- Serve as an integral part of the infantry team by identifying, fielding, and developing training and employment techniques for equipment on behalf of the entire community.

This page intentionally left blank.

Chapter 3

Mission Command

The Ranger force can work unilaterally under a JSOTF, as a regimental SOTF, or as an Army component of a JTF. Historically, the Ranger force has conducted training and operations while subordinate to a JSOTF.

COMMAND

3-1. USASOC exercises peacetime command of the Ranger Regiment. Because USASOC is a force provider for the combatant commanders and United States Special Operations Command (USSOCOM), it resources the regiment to ensure success in the operational environment.

3-2. During general war and contingency operations, the Chairman of the Joint Chiefs of Staff; Commander, USSOCOM; or the designated theater or joint force commander designates specific mission command of the Ranger Regiment. Command relationships depend upon METT-TC. Command and control is set where the Ranger-force capabilities can be best applied across the global or theater spectrum. Mission command is designed to be flexible so it can be tailored to the operational requirement. Once committed to an operation, the Ranger units are placed under the operational control of the responsible command. In some instances, Ranger forces may even operate directly under the authority of the senior U.S. Embassy representative.

3-3. Ranger mission command takes advantage of a wide variety of state-of-the-art, interoperable, secure communication and automation assets with extended ranges and capabilities. Command and control provides the capability to communicate with strategic command and control nodes.

3-4. The Ranger Regiment headquarters may serve as a SOTF headquarters when augmented with LNOs and additional staff personnel from other conventional Army and ARSOF units. This augmentation may include MIS, CA, aviation, communications, and logistic units or functions. The SOTF normally is the Army special operations component of a JSOTF. If the Ranger Regiment and an SF headquarters are under the same JSOTF commander, the JSOTF commander normally forms two or more SOTFs.

CONTROL

3-5. The Darby Network serves as the regiment's primary command and control system between its dispersed locations. The network consists of portal Web pages; mail servers; servers to inject intelligence, surveillance, and reconnaissance feeds; radio; and other network services. The portal facilitates knowledge management by easily disseminating the same information to users at all levels within the regiment. It also serves as the display and interface system to provide immediate operational and intelligence updates to streamline the targeting process and facilitate rapid prosecution of targets. The Darby Network enhances parallel planning to significantly reduce planning time and condense the planning process. The result is more time for subordinate unit planning, rehearsals, and mission preparation. Each staff member carries a laptop computer with a removable classified hard drive to an intermediate staging base (ISB) or other facility. The staff members' laptops are plugged into the Darby Network through the main command post, while the subordinate Ranger battalions link into the regiment's network. In addition, the regiment's network can be up-linked to the higher-headquarters network. This networking capability greatly facilitates real-time planning, efficiency, and timeliness during the military decision-making process phase of the operation, and battle tracking during the current operations phase.

3-6. The regimental headquarters tailors flexible command and control according to METT-TC. Options may include maintaining an HMOC and establishing a main command post, two tactical command posts, liaison teams, a Ranger deployable planning team, and cross-functional teams.

MAIN COMMAND POST

3-7. The regiment's main command post includes not only the staff, but also sets of deployable equipment and computer hardware tailored to the mission. The main command post deploys and occupies either a hardened facility, or emplaces tents with electricity and connectivity. The staff conducts the military decision-making process, provides control for current operations, and plans future operations from the main command post. The main command post—

- Analyzes assigned missions, concentrating on planning 48 to 72 hours in the future.
- Plans, coordinates, integrates, and synchronizes future operations.
- Monitors current operations and controls simultaneous operations for up to two different theaters or areas of operation.
- Provides—
  - An intelligence analysis center for Ranger forces with available links to national and theater-level intelligence architectures.
  - Control systems using the latest hardware and software applications.
  - Access to the Global Information Grid (GIG), any joint information automation networks, tactical local area networks, and the Internet.
  - Information operations planners to interface with JSOTF or JTF information operations elements.
  - Administrative, medical, CBRN, and logistical planning and coordination support to the committed Ranger force.
- Receives and analyzes real-time information and produces intelligence products for preparation of the operational environment.
- Coordinates logistical support for the force in a continental United States (CONUS) or outside the continental United States (OCONUS) ISB.
- Contains the tactical operations center in order to command and control current operations.

TACTICAL COMMAND POSTS

3-8. The regiment battle staff can deploy two tactical command posts simultaneously from the ISB into target areas to provide command and control during current operations. These command posts—

- Perform the functions of the main command post for current operations while employed in a tactical command and control configuration on the ground, aboard an airborne command and control platform (fixed- or rotary-wing), or aboard an afloat command and control platform such as an aircraft carrier or other designated command and vessel.
- Infiltrate with, or apart from, subordinate Ranger forces using foot, ground-mobility, fixed- or rotary-wing aircraft, or parachute-assault infiltration.

LIAISON TEAMS

3-9. The Ranger Regiment is capable of deploying up to four liaison teams that provide—

- Ranger representation to a higher employing staff, whether a JSOTF, JTF, theater special operations command, or theater staff.
- An operations interface to the higher employing headquarters.
- The logistics interface and required coordination.
- Fire support planning and coordination.
- Intelligence analysis and dissemination, and coordination of intelligence requirements with the employing staff.
- Communications (voice and data) between the higher staff and the employed Ranger force.
- Interfaces with all headquarters using the most current hardware and software automation applications.

RANGER DEPLOYABLE PLANNING TEAM

3-10. The Ranger deployable planning team is prepared to deploy on short notice to any combatant commander, JTF, or JSOTF headquarters (Table 3-1). The Ranger deployable planning team coordinates potential Ranger operations in support of emerging or ongoing contingency operations. The Ranger deployable planning team is composed of experienced Ranger planners and is prepared to deploy to the crisis-planning headquarters, fully equipped with planning and briefing materials. Standing operating procedures allow pushing forward team members' weapons, load-carrying systems, and rucksacks to the crisis-planning site when needed.

Table 3-1. Ranger deployable planning team composition

Position	Primary	Alternate
Team Officer in Charge (OIC)/ operations planner	Senior LNO	Ranger S-3
Intelligence planner	Assistant intelligence officer (AS-2)	Target intelligence officer
Fire support planner	Target officer	Marine exchange officer (MAREXO)
Logistics planner	Assistant S-4	Ranger S-4
Communications planner	Ranger S-6	Communications Sergeant Major

3-11. The Ranger deployable planning team deploys during a quickly developing crisis upon USASOC- or USSOCOM-approved request by a combatant commander or special operations commander. An OCONUS deployment requires a Joint Chiefs of Staff deployment order. The Ranger deployable planning team may rely on the requesting headquarters for secure communications, but the team plans only the Ranger operations.

PROVISIONAL COMMAND AND CONTROL TEAMS

3-12. The Ranger task force can organize provisional command and control teams (also known as cross-functional teams) that are assigned a target, target set, or area of operations, and are task-organized to conduct find, fix, finish, exploit, and assess operations. The cross-functional team fuses intelligence and operations, coordinates efforts with adjacent units, develops targets through reconnaissance and nonorganic organizations, and controls direct-action operations. During recent global operations against terrorist networks, the cross-functional teams became critical in synchronizing actions and generating unity of effort against specific target sets.

3-13. The cross-functional team is task-organized according to METT-TC with a command and control team that has both mission command and planning responsibilities. The cross-functional team serves as an intermediate headquarters between company-sized elements and the battalion.

3-14. The team includes the following:
- A commander.
- An operations officer.
- An intelligence analyst.
- A communications specialist.
- Joint forces, as required.

3-15. The team can be augmented with—
- A fire support officer.
- An air officer.
- A SOF unit liaison.
- A conventional force unit liaison.
- Interrogators.

This page intentionally left blank.

Chapter 4

Communications

Rangers must be able to communicate worldwide and at any time by using national, theater, and SOF communications assets. To enable assigned forces to perform the ARSOF core activities, USSOCOM developed doctrinal communications principles and architectural tenets. These principles and tenets guide communications support of special operations.

DOCTRINAL COMMUNICATIONS PRINCIPLES

4-1. Communications systems support to ARSOF must be—

- *Global*. SOF communications systems span the full range of diverse special operation missions worldwide. SOF communications make maximum use of existing national capabilities and commercial, tactical, and host-nation assets. Access to the infosphere is available at the lowest possible tactical level.
- *Secure*. Employment of SOF communications systems involve the use of the latest technology procedures and National Security Agency–approved encryption and devices that prevent exploitation by the enemy.
- *Mission-tailored*. SOF communications systems deploy relative to the projected operational environment, information-transfer requirements, threat, and mission analysis.
- *Value-added*. SOF communications systems never compromise a unit on the ground, in the air, or at sea. Flexibility and interoperability of communications systems substantially increase the fighting effectiveness of the SOF warrior.
- *Jointly interoperable*. SOF communications systems are interoperable by design, adapting to varying control structures. They support operations with joint, multinational, and interagency forces.

ARCHITECTURAL TENETS

4-2. Doctrinal principles and planning considerations are the building blocks for an operational architecture that guides communications systems strategy. The architectural tenets for ARSOF communications systems provide SOF operators seamless, robust, protected, and automated communications systems by using the full frequency spectrum. In addition, the tenets also ensure the systems comply with established standards. Implementing these tenets eliminates traditional geographical, procedural, and technical boundaries. The infosphere allows SOF elements to operate with any force combination in multiple environments. The architectural tenets are—

- *Seamless*. Digital SOF communications systems are transparent to the warrior and support every phase of the mission profile—in garrison, in transit, and while deployed. Multiple entry points into the infosphere, high-speed networks, and worldwide GIG connectivity are critical elements of this tenet.
- *Robust*. Robust networks feature multiple routing, alternative sources of connectivity, bandwidth on demand, and modularity and scalability. Using multiple routing and alternative connectivity sources prevents single points of failure and site isolation. Bandwidth on demand and automatic network reconfiguration is available for garrison locations via secure commercial means, as well as via the Single Channel Antijam Manpack Portable Interface (SCAMPI) entry points for tactical SOF.
- *Protected*. SOF control nodes are lucrative targets for all types of adversaries, ranging from foreign governments to the hacker. A thorough information assurance plan protects SOF

communications architectures and critical resources from attack and intrusion. Creating and maintaining sound communications security (COMSEC), computer security, and information security programs are necessary to protect the network.

● *Automated.* SOF communications systems must facilitate the exchange of digital data and implement advanced automation techniques to reduce operator manning and to exploit unattended operation. Full automation facilitates the exchange of information with all operators in the mission, including elements of a multinational joint force, other SOF components, a wide range of intelligence sources, and national information sources. Networking technology of the local area network and the wide area network is the cornerstone of a digital, seamless nodal architecture that provides transparent connectivity at all echelons.

● *Full spectrum.* SOF communications systems are not limited in using the entire frequency spectrum for information transfer. SOF has been at the forefront of exploiting new technology to take advantage of more of the frequency spectrum. In addition, blending in available host-nation assets can enhance many of the special operations missions. ARSOF must be in a position to not only use emerging technology but also technology outside traditional high frequency (HF) and man-portable satellite systems.

● *Standards compliant.* SOF communications systems adhere to commercial, international, federal and Department of Defense hardware and software standards. Adhering to Department of Defense communications standards ensures the capability of interchanging hardware and software products. Adherence also permits the interface and exchange of data with all organizations that support or require SOF through the infosphere and GIG.

COMMUNICATIONS ELEMENTS

4-3. The Ranger Regiment has an S-6 located in the regimental headquarters and a regimental communications company located in the regimental special troops battalion. Each Ranger battalion has an S-6 and a battalion signal section located in the battalion HHC.

REGIMENTAL S-6

4-4. The S-6 is the principal staff officer for all matters concerning signal operations, automation management, network management, and information security. The staff officer plans signal operations, prepares the signal annex to the operations orders, and recommends employment of Ranger communications assets. Additionally, the staff officer ensures that redundant signal means are available to pass time-sensitive command information from collectors to processors. The S-6, S-2, and S-3, support electronic warfare operations focusing on electronic protection. The S-6 also supervises the regiment's frequency manager and COMSEC custodian.

REGIMENTAL COMMUNICATIONS COMPANY

4-5. Although the regimental communications company is designed to support two teams, it is task-organized into three teams. Each team provides communications for a separate tactical operations center. The regimental communications company includes an electronic maintenance section.

BATTALION SIGNAL ELEMENTS

4-6. The battalion S-6 is the principal staff officer for all matters concerning signal operations, automation and network management, and information security at the battalion level. Battalion S-6 duties are similar to the regimental S-6. The battalion signal section is task-organized into three teams, with each team capable of supporting separate tactical operations centers.

COMMUNICATIONS CAPABILITIES AND EQUIPMENT

4-7. Single channel ultrahigh frequency (UHF) satellite communications (SATCOM) make up the backbone of the Ranger communications for links between each deployed headquarters. However, with technology continually progressing, other capabilities better enable the commander and his staff to maintain information superiority. Rangers have low-data-rate capabilities through a narrowband data controller. This data controller links computers via single channel UHF tactical satellite (TACSAT) or HF automatic link establishment (ALE) systems. Some of the other low-data-rate systems include systems based upon international maritime satellite (INMARSAT) technology, including the SOF-deployable node-light. The Rangers are provided better bandwidth access which enables them to maintain the common operational picture (COP) through the following systems:

- SATCOM systems capable of medium data rate.
- SOF-deployable node medium system.
- Joint medium data system.

4-8. The Ranger Regiment is not provided with the organic high-capacity multichannel SATCOM. Therefore, the joint communications support element or the 112th Signal Battalion (Airborne) must augment the Ranger Regiment when needed. Figure 4-1 shows the architecture of deployed Ranger communications. Other single-channel tactical capabilities include intrateam radios and HF-ALE radio systems.

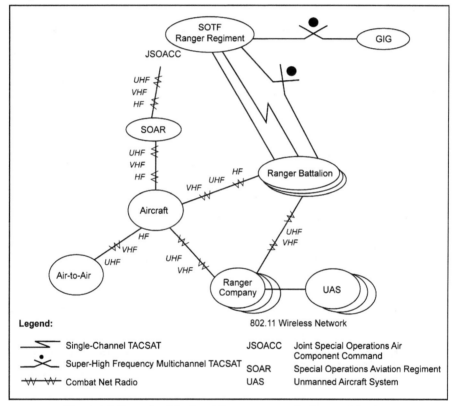

Figure 4-1. Example of deployed Ranger communications architecture

4-9. The Ranger Regiment provides specialized communications training for all 25-series, electronic maintenance, platoon radiotelephone operators and forward observers. The training includes:

- Every radio system within the regiment.
- User-level automation support.
- Vehicle communications platforms.
- INMARSAT-based systems.
- Non-line-of-sight radio repeaters.
- Specialized circuits (such as SECNET-11 encrypted wireless networking).
- Intelligence, surveillance, and reconnaissance video feeds.

A Ranger normally will complete the month-long course following the Ranger Assessment and Selection Program 1. The goal of the training is to provide every platoon leader and platoon sergeant with a trained radiotelephone operator who understands all the communications systems within their unit.

CONCEPT OF EMPLOYMENT

4-10. The mission of the Ranger Regiment has a global rather than a regional orientation. It is not designed for long-term employment. Ranger communications must be rapid and able to support airborne, air-assault, and infantry-type operations at all echelons. When supporting conventional forces, the Ranger Regiment provides a Ranger liaison element to the supported headquarters.

4-11. Rangers rely heavily on external fire support. Ranger fire support personnel train extensively on the employment of close-air support, attack helicopters, naval gunfire, AC-130 gunships, and artillery. Redundant and reliable communications to these fire support platforms are essential. The SOF joint fires element further enables a SOTF, including Rangers, to better use fire support assets while reducing friendly and civilian fratricide.

COMMUNICATIONS SUPPORT

4-12. The Rangers use secure, long-range, lightweight, real-time, HF and SATCOM in support of Ranger operations. Effective long-range communications provide control links between deployed Ranger units and the controlling headquarters. The JTF, Army forces, or theater commander is responsible for communications between the controlling headquarters and the Ranger Regiment. The SOF commander may also provide secure communications terminals to the Ranger Regiment or a deployed battalion.

4-13. Secure amplitude modulation (AM), frequency modulation (FM), and SATCOM radios are the primary means of communication within the Ranger Regiment. Within the Ranger battalion, AM/FM/UHF radios provide communications to company, platoon, squad, and individual Soldier levels.

4-14. During certain missions, specially trained and equipped quick-reaction elements deploy with the Ranger force to provide secure communication links to the SOF commander. The quick-reaction elements operate on either SATCOM or TACSAT channels or through an airborne communications relay platform. Depending on the mission, an airborne battlefield command and control center aircraft, a joint airborne communications center, or a command post may be used. Their communications systems can operate at all levels of the national chain of command to permit a quick response to the tasking authority.

4-15. When only one Ranger battalion is employed and the regimental headquarters is not the controlling headquarters, the Ranger Regiment provides a liaison cell to the controlling higher headquarters. This liaison cell includes a communications element from the regimental communications company. It can provide secure SATCOM voice and data support to the Ranger force in the objective area.

4-16. Generally, if two or more Ranger battalions are employed, the Ranger Regimental headquarters deploys and acts as the control headquarters. The regimental communications company would then provide another communications link to the SOF commander.

4-17. The Ranger Regimental S-6 ensures that the necessary communications links are set up and coordinated. The many communication means and channels available provide an effective control of a deployed Ranger force. However, communication means must be closely coordinated at all levels to control the complex operations of a Ranger mission. The planning and coordination with supporting aviation, transportation, fire support, medical, and logistical elements prior to an operation are vital to efficient communications. The regimental S-6 must consider the communication systems linking the Ranger force and other Services. Air-to-ground and ship-to-shore communications are vital and must be set up early in Ranger operations.

4-18. The Ranger Regiment has a substantial communications capability that allows it to plug into the communications systems of any potential employing headquarters. The regimental communications company, with oversight from the regimental S-6, is responsible for the tactical communications architecture within the regiment. The regiment also has an automated data processing section that is responsible for maintaining the tactical web planning system, the USASOC local area network, and other classified and unclassified local area networks. Additionally, it is responsible for digitizing Ranger standing operating procedures, manuals, policies, and routine paperwork. The regimental communications company–

- Provides—
  - Secure UHF, SATCOM, FM, VHF, and tactical local area network capabilities at the ISB.
  - Secure UHF, SATCOM, FM, VHF, and tactical automation to the regimental headquarters and Ranger battalions in the area of operation.
  - Multichannel SHF commercial satellite capability.
  - High-bandwidth, microwave line-of-sight capability.
  - Access to employing headquarters local area networks and the Internet.
  - Secure in-flight communications on rotary-wing aircraft and enroute communications on fixed-wing aircraft using both voice and data communications.
  - Cryptological support to the regimental headquarters.
- Operates from the ground out of a rucksack or from vehicular control or communications platforms.
- Conducts organizational, direct support and general support repair of radio, cryptological, and computer equipment for the regimental headquarters.
- Establishes a tactical local area network and access to local area network systems that connect to higher operational headquarters, USASOC, and USSOCOM, as well as SECRET Internet Protocol Router Network access.

4-19. The regimental S-6 maintains an automated data processing section that—

- Manages all aspects of automated data processing hardware tactical repair, replacement, and training.
- Provides and maintains the regimental tactical web.
- Conducts life-cycle replacement program and supervises the evacuation and or replacement of defective communications and electronics equipment.
- Provides automation repair and user operator instructions.

This page intentionally left blank.

Chapter 5

Intelligence

Normally as part of an integrated SOF contingent, Rangers provide a responsive strike force for conducting direct-action missions. Ranger operations rely on the elements of surprise, precise planning, and orchestrated execution to conduct special missions supporting vital U.S. interests. Intelligence support for Ranger operations primarily focuses on providing target-specific information for objectives of strategic or national importance. In a Ranger battalion, the intelligence structure parallels that of a light infantry battalion; however, the intelligence structure within a Ranger battalion is more robust. The organic intelligence structure within the Ranger Regimental headquarters is larger than the intelligence structure of an infantry brigade, and is structured to provide linkage from higher echelons to the operational units. Additionally, the Ranger Regiment has an organic reconnaissance capacity in the form of the regimental reconnaissance company.

ROLES

5-1. The Ranger Regiment's task is to plan and conduct special operations against strategic or operational targets in pursuit of national or theater objectives (FM 3-05). The Ranger Regiment and its subordinate battalions have a worldwide focus. Ranger missions are diverse and carried out on any terrain and under any condition. Ranger operations have high-risk and high-payoff attributes. Therefore, accurate, detailed, and timely intelligence is critical for planning and executing Ranger missions. Because the regiment has limited organic intelligence assets, it is through active interface with the supporting intelligence system that Rangers receive answers to specific target intelligence requirements.

5-2. Ranger direct-action operations may support or be supported by other special operations activities, or they may be conducted independently or with conventional military operations. Ranger direct-action operations typically include—

- Raids against targets of strategic or operational value.
- Lodgment operations (such as airfield seizures).
- Noncombatant evacuation operations (NEOs).

ARMY SPECIAL OPERATIONS FORCES INTELLIGENCE CRITERIA

5-3. ARSOF missions are intelligence driven and intelligence dependent. Intelligence products developed for ARSOF must be detailed, accurate, relevant, and timely. For example, infiltrating a building in a nonpermissive NEO requires exact information on its structure and precise locations of hostages or persons to be rescued. National-level and theater-level intelligence products are often required at a lower echelon than is normally associated with support to conventional forces. They also may require near-real-time dissemination to the operator level.

5-4. ARSOF intelligence requirements are heavily dependent on the specific mission and the current situation. Because ARSOF missions may vary widely, the associated intelligence support requirements also may vary. Therefore, intelligence production for SOF requires a thorough understanding of special operations requirements at the tactical level. Intelligence production presents national and theater intelligence producers with unusual production and dissemination challenges. The following variables can affect intelligence infrastructure requirements:

- Combat (nonpermissive) or cooperative noncombat (permissive) environments.
- Multinational, combined, joint, or unilateral operations.

- Force composition.
- Maritime- or land-based operations.
- Mission duration.
- Command and control elements and intelligence support facilities.
- Adversary capabilities, objectives, and operational concepts.

INTELLIGENCE CRITERIA FOR DIRECT ACTION MISSIONS

5-5. Intelligence criteria supports direct action, special reconnaissance, and counterterrorism missions. Because SOF missions applying direct military force concentrate on attacking or collecting information on critical targets, the information required is highly perishable, requires near-real-time reporting, and often requires special handling to protect sources.

5-6. Rangers engaged in these missions depend on detailed and current target materials for mission planning and execution. Rangers require extensive information from national, theater, and SOF-specific order of battle (OB), threat installation and target assessment databases, files, studies, and open-source information. Rangers require current intelligence updates on targets and target changes from assignment of the mission through planning, rehearsal, execution, and post strike evaluation.

5-7. The basis for successful Ranger mission planning is the target intelligence package (TIP) normally developed by the theater joint intelligence center or joint analysis center (United States European Command only). TIPs must contain timely, detailed, tailored, and fused multisource information describing—

- Target description.
- Climate, geography, or hydrography.
- Demographic, cultural, political, and social features of the joint special operations area.
- The threat, including the strategy and force disposition of the military, paramilitary, or other indigenous forces, as well as any forces that endanger U.S. elements.
- Infiltration and exfiltration routes.
- Key target components, including line of communications (LOCs).
- Threat command, control, and communications.
- Threat information systems.
- Evasion and recovery information.

5-8. Current geospatial (imagery, mapping, and geodesy) products of the target and area of operation are an important part of any TIP. Ranger elements in pre-mission isolation use TIPs as primary intelligence resources. The TIPs help focus requests for intelligence information not covered or for data requiring further detail.

5-9. During all phases of these missions, Rangers depend upon the timely reporting of detailed and highly perishable current intelligence related to their operational situation. They also require rapid real-time or near-real-time receipt of threat warnings to enable them to react to changing situations and to ensure operational area security.

INTELLIGENCE REQUIREMENTS

5-10. Intelligence support requirements for the Ranger Regiment center on target intelligence. Ranger forces conducting direct-action operations are employed against targets of strategic or operational value. These targets may be similar in nature and location to targets being interdicted by conventional means— such as a weapons of mass destruction facility—but because of the type of unit being employed, a higher level of intelligence resolution is required. For instance, a weapons of mass destruction facility being targeted for interdiction through the air tasking order may require using only signals intelligence (SIGINT), measurement and signature intelligence, and geospatial intelligence. A target being interdicted by a Ranger direct-action mission may require products from these disciplines as well as an extensive use of human intelligence to bring the level of detail from the general exterior of the facility and a few salient physical features to a detailed product encompassing interior details, guard force composition, schedules, protocols, and command and control. During a NEO, commanders must possess a detailed picture of the human terrain. Intelligence preparation for a NEO typically requires detailed data and intelligence and electronic warfare support from national- or joint-level agencies.

INTELLIGENCE ORGANIZATION

5-11. Intelligence assets organic to the Ranger Regiment are organized according to operational and analytical needs. The Ranger Regiment has organic assets it can use to perform intelligence functions and missions. The regimental S-2 section and the military intelligence company are structured to support the—

- SOTF with a robust, sustained operational capability.
- STRIKE force with short duration capability in support of national missions.
- HSOC with a sustained garrison capability.

REGIMENTAL S-2 SECTION

5-12. The regimental S-2 section is shown in Figure 5-1. As the primary intelligence advisor to the commander, the S-2—

- Oversees all intelligence operations for the 75th Ranger Regiment.
- Provides the commander with all-source intelligence assessments and estimates at the tactical, operational, and strategic levels.
- Directs tasking of intelligence collection assets, manages interrogation operations, interprets imagery from overhead and other systems, directs counterintelligence and support to operational security operations, and manages SIGINT operations, including jamming and participation in deception operations.
- Identifies, confirms, and coordinates area requirements for geospatial information and services products to support operation plans and concept plans.

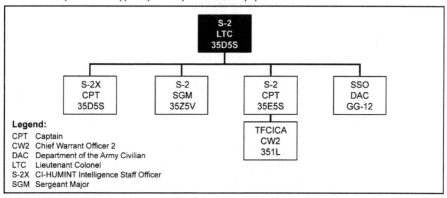

Figure 5-1. Regimental S-2

5-13. The S-2X assists the S-2 with all of the above responsibilities as well as being the senior military intelligence representative in the regimental alternate assault command post. Additionally, the S-2X, when required, serves as the intelligence LNO to other special operations units.

5-14. CI/HUMINT operations are managed by the S-2X. The S-2X is responsible for directing, coordinating, and advising on all CI/HUMINT operations within the regimental area of operations. The S-2X supervises the execution of the regimental CI/HUMINT collection plan, to include liaison and deconfliction with adjacent units in support of current operations. The S-2X also interfaces with higher headquarters and outside agencies to ensure adequate CI/HUMINT support to the regiment.

5-15. The special security officer ensures that sensitive compartmented information is properly secured, maintained, and accredited.

REGIMENTAL MILITARY INTELLIGENCE COMPANY

5-16. The military intelligence company (Figure 5-2) is the principal source of intelligence support to Ranger commanders. The military intelligence company capabilities include provision of—

- Sustained multidisciplined intelligence support to the SOTF.
- Ground-based SIGINT operations.
- Sustained CI/HUMINT operations in support of SOTF operations.
- Operation of facilities for enemy prisoners of war and the conduct of interrogations.
- Detailed imagery, geospatial, and terrain products.
- Analytical reach-back from HSOC.

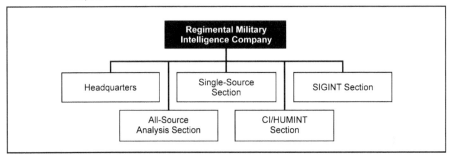

Figure 5-2. Regimental Military Intelligence Company organization

ALL-SOURCE ANALYSIS SECTION

5-17. The all-source analysis section (Figure 5-3) uses various systems and software to graphically display the enemy situation and produce target folders for operations. The section consists of three OB teams and one HSOC intelligence support team. The all-source analysis section also maintains databases of intelligence information, conducts pattern analysis, exploits national databases for needed intelligence, and searches for relevant message traffic to build and display the enemy picture. This section produces intelligence products tailored for dissemination down to company level to facilitate operator tactical usability.

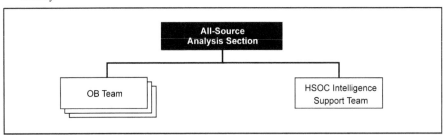

Figure 5-3. All-source analysis section

SINGLE-SOURCE SECTION

5-18. The single-source section (Figure 5-4) comprises three subordinate sections responsible for individual intelligence disciplines. These include the following:

- *Imagery analysis section.* The imagery analysis section is responsible for the acquisition, exploitation, and dissemination of imagery, motion video, and geospatial data and products from national and tactical assets in direct support of Ranger missions. Imagery analysts are also responsible for battle damage assessment analysis using current imagery. The section is augmented with a civilian imagery analyst from the National Geospatial-Intelligence Agency.

- *Terrain analysis section.* The terrain analysis section, along with a civilian geospatial analyst provided by the National Geospatial-Intelligence Agency, performs cartographic and terrain analysis. This section provides the ground maneuver units with detailed analysis of terrain, to include LOCs, terrain elevation, and natural and manmade obstacles. The terrain analysis section predicts terrain and weather effects as applied to mission command, communication, and computer and intelligence systems.

- *Technical Control and Analysis Element (TCAE).* The TCAE performs technical analysis. SIGINT analysts in the TCAE operate SIGINT–related programs accessed though the Joint Worldwide Intelligence Communications System and the National Security Agency network. The TCAE analyst uses a combination of Army and special operations–provided intelligence systems. These systems are designed to be interoperable with theater intelligence systems and national assets, such as the Tactical Exploitation of National Capabilities Program.

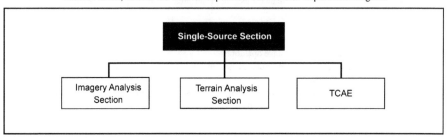

Figure 5-4. Single-source section

COLLECTION MANAGEMENT AND DISSEMINATION SECTION

5-19. The collection management and dissemination section is part of the headquarters element. It comprises an officer and an NCO. The section synchronizes and monitors the S-2's collection requirements with internal and external collection sources. The collection manager plans and coordinates strategic surveillance and reconnaissance requirements. Duties also include conducting research and exploitation of national databases using specific tools to ensure an all-source approach to analysis and most current data used for TIPs.

COUNTERINTELLIGENCE/HUMAN INTELLIGENCE SECTION

5-20. The CI/HUMINT section includes one counterintelligence team and two HUMINT collection teams (HCTs) that are commanded and controlled by the operational management team. The operational management team is responsible for the overall management of all CI/HUMINT teams operating in the regimental area of responsibility. Properly trained CI/HUMINT members of CI and HUMINT teams conduct Military Source Operations (MSO) to answer regimental intelligence requirements. Additionally, CI teams conduct CI collection for force protection and counterintelligence requirements. The HCT has the capability to conduct interrogations of detainees.

SIGNALS INTELLIGENCE SECTION

5-21. The SIGINT section comprises three operational SIGINT teams that conduct ground-based SIGINT collection to—

- Find and fix the location of the enemy.
- Provide indications and early warnings of enemy intent or actions.
- Provide SIGINT force protection.

5-22. The SIGINT analysis section gathers, sorts, and analyzes intercepted messages from the operational SIGINT team to isolate valid message traffic. Additionally, they evaluate intelligence data from SIGINT reports released from national agencies. The information gathered is used to identify actionable targets.

REGIMENTAL RECONNAISSANCE COMPANY

5-23. The Regimental Reconnaissance Company (Figure 5-5) comprises six Ranger reconnaissance teams, a selection and training team, an operations section, and a headquarters element. The Regimental Reconnaissance Company's primary mission is to conduct all forms of reconnaissance and surveillance and limited direct action to support JTF missions. Regimental Reconnaissance Company teams can operate directly for a Ranger battalion commander or in support of the regiment or higher headquarters. The Regimental Reconnaissance Company teams give the regiment the capability to conduct shaping operations and answer intelligence requirements. The Regimental Reconnaissance Company—

- Fulfills target area surveillance missions in an area before committing Ranger or other SOF elements to the operation.
- Engages hostile targets with direct fire, indirect fire, and demolitions.
- Conducts—
  - Limited terminal guidance operations.
  - Prestrike and post-strike surveillance on critical nodes for battle damage assessment direct action requirements.
  - Pathfinder operations to reconnoiter, select, clear, and prepare landing zones and drop zones.
  - Autonomous tactical operations for up to 5 days in denied areas.
- Provides information on threat OB and target sites and conducts route and limited CBRN reconnaissance.

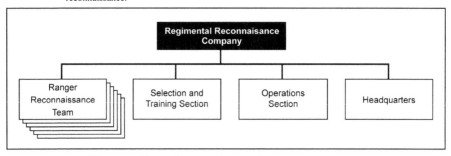

Figure 5-5. Ranger Reconnaissance Company organization

5-24. The Regimental Reconnaissance Company conducts infiltrations and exfiltrations using several methods including—

- Military free fall, including high-altitude, low-opening and high-altitude, high-opening.
- Vehicles (standard/nonstandard).
- Low-visibility actions.
- Small boats.
- Scout swimmers.

BATTALION S-2 SECTION

5-25. The Ranger battalion S-2 section (Figure 5-6) comprises three officers—the S-2 and two AS-2s—and five enlisted Soldiers. The section has a limited capability to collect and analyze information. Its mission is to support the battalion commander with basic intelligence, database maintenance, collection management, analysis, and tactical intelligence production and dissemination for battalion operations. The military intelligence company or external assets, as required, augment the battalion S-2 section. Additionally, the Ranger battalion S-2—

- Tasks, through the S-3, battalion elements to perform combat intelligence missions supporting battalion operations.
- Conducts intelligence training for battalion elements.
- Supports the planning, coordination, and execution of Ranger target rehearsals.
- Briefs and debriefs reconnaissance teams.
- Identifies, confirms, and coordinates priorities for input to regimental geographic area requirements for geospatial information and services products.
- Maintains responsibility of personnel and physical security.

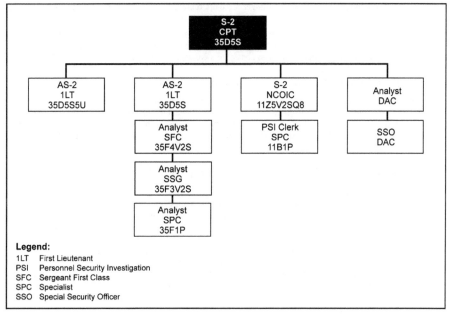

Figure 5-6. Ranger battalion S-2 organization

OTHER ORGANIC SUPPORT

5-26. The regimental and battalion surgeons are a source of medical and CBRN intelligence concerning possible deployment locations. They also provide valuable information on disease and health conditions in the area of operations. Fire support personnel are able to serve as a conduit for information collected by units providing supporting fires. This may include detections by the counterbattery radar of an artillery unit or observations by pilots flying close air support missions.

NONORGANIC INTELLIGENCE SUPPORT

5-27. Intelligence support for Ranger operations is specialized and sensitive. Nonorganic intelligence support to the Ranger Regiment is discussed in the following paragraphs.

SUPPORT FROM HIGHER HEADQUARTERS

5-28. The USASOC Deputy Chief of Staff for Intelligence (G-2), Intelligence Operations Division, coordinates with USSOCOM Special Operations Intelligence and Information Operations, Special Operations Command joint intelligence center, and theater- and national-level intelligence agencies for the intelligence needs of the Ranger Regiment.

5-29. When deployed, the JSOTF commander provides intelligence support to the Ranger Regiment. The JSOTF intelligence directorate (J-2) provides target-specific intelligence in the form of a TIP. The JSOTF J-2 also provides multidiscipline counterintelligence and electronic warfare support to the Ranger force. The coordination among the regiment, the corps or echelons above corps intelligence node, and the targeting center is essential to effectively employ Ranger forces. The Ranger Regiment normally places an intelligence LNO at the JSOTF or the appropriate intelligence production facility to ensure—

- The intelligence needs of the Ranger force are relayed to the appropriate intelligence-processing center.
- The resulting analysis is based on the specific needs of the Ranger force commander.
- The coordination is made with the corps or echelons above corps targeting center when the Ranger force is under operational control to a conventional force.

INTELLIGENCE, SURVEILLANCE, AND RECONNAISSANCE SUPPORT

5-30. When deployed, additional intelligence, surveillance, and reconnaissance assets are provided by the JSOTF to assist the military intelligence company in detection and target refinements. Although the Ranger Regiment has a tactical UAS program, overhead full-motion video by national assets, such as the Predator UAS, normally is provided by the JSOTF. Additional collection assets normally requested through the JSOTF are national SIGINT collection and national imagery collection.

RANGER SUPPORT TO THE INTELLIGENCE PROCESS

5-31. The robust capability of the Regimental Reconnaissance Company to conduct reconnaissance through its six Ranger reconnaissance teams not only provides the regiment with organic intelligence, surveillance, and reconnaissance capability, but complements the capability of the force in theater. Rangers often are the first to encounter the enemy and can confirm or deny friendly assessments of threat organization, equipment, capabilities, and morale. They can bring back captured threat equipment for evaluation and report on the effectiveness of friendly weapons on threat systems. Rangers also can provide real-time assessments of the target area civilian population's morale and their physical disposition for use in military information support operations (MISO) and CA plans for follow-on forces. Ranger S-2s must be proactive in debriefing to ensure this valuable information enters the intelligence cycle.

Chapter 6

Employment, Movement, and Maneuver

The Ranger Regiment conducts direct-action operations by order of the President or Secretary of Defense and according to the warfighting commanders' national military strategy. Ranger forces can be used in major regional contingencies, provide credible overseas presence and regional engagement, participate in multinational peace operations, conduct punitive attacks, support U.S. counternarcotics operations, and protect the lives and security of American citizens abroad. Rangers provide the joint force with a unit of unmatched expertise in airfield seizure and raid capability.

GENERAL

6-1. Rangers are apportioned by the Joint Strategic Capabilities Plan to theater warfighting commanders and are integrated into the Chairman of the Joint Chiefs of Staff and theater concept plans. The Ranger force normally is employed as part of a JSOTF. It provides a worldwide, strategically responsive strike force with a highly lethal ground combat capability. The Ranger force can serve as a flexible deterrent option to demonstrate U.S. national resolve by immediately committing military power on land into a threatened area. It can also conduct offensive, direct-action operations against targets of strategic or operational value to achieve theater campaign or major operational objectives.

6-2. Ranger forces are used when a rapid-response, highly disciplined, and lethal infantry force is required. Rangers can conduct forcible entry, capture or destroy targets of operational or strategic significance, or perform rescue and recovery operations. Ranger forces conduct these missions in politically sensitive environments, under intense media scrutiny, and when the President, Secretary of Defense, or Chairman of the Joint Chiefs of Staff requires the highest probability of success. Ranger forces conducting direct-action operations may—

- Seize, destroy, or capture enemy forces or facilities.
- Perform reconnaissance to assist conventional commanders, special operations units, and nonmilitary organizations in finding and fixing targets.
- Recover designated personnel (conduct NEOs, liberate friendly prisoners of war, capture designated enemy personnel) or equipment in hostile, denied, or politically sensitive areas.
- Exploit sensitive material and targets to locate follow-on objectives.

6-3. Ranger forces conduct direct-action operations independently or in support of a campaign plan. Direct-action operations may be conducted in coordination with conventional forces, but differ from conventional operations in degree of risk, operational techniques, and modes of employment. Ranger direct-action operations rely on—

- Multiple insertion means (rotary-wing aircraft, ground mobility assets, parachute, boat, and foot) to maximize combat power at the decisive point.
- Undetected insertion and rapid movement to the target if the force is inserted offset from the objective.
- Surprise and shock if the insertion is on the target.

6-4. Ranger operations will, on occasion, be conducted before political preparation of the operational environment has been completed or initiated. Rangers operate under the widest range of combat conditions.

6-5. The strategic and operational reach of the Ranger force provides the President and Secretary of Defense a credible combat capability for protecting selected vital U.S. interests and citizens without having to wait for international support or guarantees of nonintervention. The Ranger force frequently is the principal element of ground combat power when the U.S. conducts a forcible entry operation.

DEPLOYMENT OPTIONS

6-6.   The 75th Ranger Regiment can deploy a force within 18 hours of notification. It can follow with additional forces within 72 hours. The regiment's headquarters maintains command and control, liaison, communications, and reconnaissance elements immediately available for deployment. A higher status of readiness in response to specific world situations can be achieved. Options include deploying—

- Directly from the home station to the target.
- From the home station to a CONUS or OCONUS ISB with logistical unit support; then launching to a target, a forward staging base (FSB), or mission support base.
- From the home station to a seaport of embarkation to embark on a naval vessel, such as an aircraft carrier or other suitable vessel that serves as an afloat FSB. The vessel deploys to the area of conflict. The Ranger force, along with embarked special operations rotary-wing aircraft, conduct air-assault operations launching from the afloat FSB and striking littoral targets.

6-7.   In the recent past, Rangers have conducted mostly short-duration strike operations with platoon- to battalion-sized elements—

- From an ISB, SOTF, or mission support site (MSS).
- To seize terrain or destroy facilities.
- Based on intelligence provided by national or operational assets.
- Using airborne or air assault as the primary insertion platform.

6-8.   While operating in support of the global operations against terrorist networks, Rangers conduct sustained, full-spectrum operations of a battalion, company, platoon, and section size—

- From a SOTF or MSS.
- In the operational environment of conventional units.
- To kill or capture high-value targets or high-value individuals.
- Based on intelligence developed through coordination with operational or national assets.
- Most often using ground mobility and rotary-wing as the primary insertion platforms.

PLANNING

6-9.   Planning for Ranger operations normally takes two forms:

- *Deliberate operations.* Deliberate operations feature meticulous contingency planning for every phase of the operation. Deceptive countermeasures and absolute secrecy, thorough preparation and rehearsals, and decisive execution characterized by surprise, precision, and audacity are accomplished before the enemy can react in strength.
- *Quick-response operations.* Quick-response operations capitalize on crisis action planning and the high combat readiness of Ranger units. Quick response is predicated on the ability to rapidly deploy and accomplish the mission before the deployment or organization of a credible threat within the target area.

Objective Serpent

On 27 March 2003, elements of the 3d Ranger Battalion conducted a night combat parachute assault to seize the H1 airfield (Objective Serpent) located in western Iraq. This operation provided one of the first forward operating bases deep within Iraqi territory. Planning for this operation found its roots in the regiment's preparation for seizing the H2 airfield (an operation that was never executed). Months of planning and rehearsals were conducted to seize the airfield, reduce significant obstacles on runways, repair runways, and expand the airhead line to receive follow-on forces.

Two full-force rehearsals were conducted weeks prior to deploying in support of Operation IRAQI FREEDOM. Numerous planning sessions and staff exercises were conducted at all levels to synchronize the operation. The detailed planning and thorough preparation for seizing the H2 airfield was easily incorporated into the mission to seize the H1 airfield.

War and Regional Contingencies

What began in the predawn hours of 4 March 2002 as an insertion to recover a Navy sea air land team (SEAL) —stranded during a compromised reconnaissance mission—quickly disintegrated into a 17-hour running firefight for one Ranger platoon. When one of their two CH-47 Chinooks was hit by an al-Qaeda rocket-propelled grenade and forced to crash-land on the Takur Ghar mountaintop in eastern Afghanistan's Shah-i-Kot Valley, the platoon found itself in the middle of murderous crossfire. With three Rangers killed in the first minutes of combat and nearly everyone injured in some way, the platoon somehow managed to exit the wrecked Chinook, establish a perimeter, tend to the seriously wounded, and return fire on a very determined enemy. The Rangers faced constant small arms and grenade attacks. While half of the Rangers in the platoon were fighting for their lives, the second Chinook found an area to set down and drop off the other half of the quick-reaction force. After they exited the helicopter, they learned they were about 2,000 feet below the pinned-down Rangers. At an altitude of over 10,000 feet and with snow varying from ankle to waist deep, the reinforcements from the second Chinook made their way up the mountain. After two hours of fighting the elements and enemy fire, the platoon linked up at the summit. Leaders organized an assault on the enemy forces, as others raced in and out of cover to help the wounded. The final assault eliminated the immediate threat, but it was then around noon and an extraction was deemed too risky until nightfall. With al-Qaeda forces in the area, a helicopter would be too tempting of a target in broad daylight. Maintaining control of the mountaintop proved a daunting challenge. For nearly eight hours, the Rangers were sporadically attacked and harassed by grenades and rifle fire. The Rangers were finally evacuated at around 2015, almost 14 hours after the first Chinook crash and almost 17 hours after the Navy SEALs first engaged the enemy.

RANGER ROLE IN NATIONAL MILITARY STRATEGY

6-10. The following paragraphs describe how Ranger forces have been used to support the national military strategy.

GLOBAL OPERATIONS AGAINST TERRORIST NETWORKS

6-11. The 75th Ranger Regiment has been deployed continuously since October 2001. Ranger missions supported Operation ENDURING FREEDOM in Afghanistan and Operation IRAQI FREEDOM in Iraq. Ranger forces conducted the following operations in support of the global operations against terrorist networks: seizure of Haditha Dam, prisoner of war rescues, airborne assaults, raids, cordon and search operations, operations with time-sensitive targets, sensitive-site exploitations, deep-look surveillances, and LOC interdictions.

REGIONAL CONTINGENCIES

6-12. When ground forces are required, the Ranger Regiment may be used as a flexible deterrent option to prevent the imminent outbreak of hostilities. If deterrence fails, Ranger forces can be the theater deep-penetration strike force of choice to raid or interdict targets of strategic or operational significance. Examples include the following:

- The 2d Ranger Battalion conducted a strategic deployment to Honduras in December 1985 to deter the Nicaraguan military who were pursuing the Contra guerrillas into Honduran territory. The Contras were using areas in Honduras along the Nicaraguan border as safe areas. The purpose of the Ranger operation was to deter continued violation of the border area and convince the Nicaraguan government to move their military back across the border into Nicaragua.
- Elements of the 1st Ranger Battalion deployed to Saudi Arabia to participate in Operation DESERT STORM. Rangers conducted an air assault raid in February 1991 to destroy an Iraqi microwave communications tower that could not be successfully destroyed by air power. This was a target of operational significance.
- The U.S. Central Command used the 1st Ranger Battalion as a flexible deterrent option following Operation DESERT STORM, which demonstrated the continuing resolve of the U.S. in the Kuwait theater of operations. The battalion conducted an airborne assault on 8 December 1991, onto Ali Al Salem Airfield in Kuwait, followed by a battalion live-fire raid exercise. The purpose was to deter Iraqi military elements from crossing the border and recovering equipment they had abandoned in Kuwait during Operation DESERT STORM.

CREDIBLE OVERSEES PRESENCE

6-13. Combatant commanders may use Rangers when a ground force is required to establish a credible American presence in any area of the world to demonstrate U.S. resolve or interest. Rangers will continue to participate in Joint Chiefs of Staff-directed exercises overseas facilitating regional engagement through military-to-military contact, joint interoperability, and adaptive joint force packaging. For example, the Ranger Regiment routinely participates in combined and JTF exercises that involve adaptive joint-force packaging. Rangers operate from naval platforms afloat with other SOF to conduct direct-action missions across the spectrum of conflict in littoral areas.

PEACE OPERATIONS

6-14. The Ranger Regiment can conduct direct-action missions in conjunction with peace operations when the controlled and disciplined use of lethal force is required. Rangers may also conduct forcible-entry operations, destruction raids, or rescue and recovery operations in conjunction with multinational peace operations. For example, in September and October of 1993, elements of the 3d Ranger Battalion conducted direct-action operations in Somalia to capture warlord Mohamed Farrah Aidid and his leadership. Aidid and his followers were interfering with the humanitarian distribution of food and supplies and attacking United Nations (UN) forces in Somalia. The purpose of the Ranger operations was to set the conditions to allow freedom of movement throughout the area of operations for UN forces.

PROTECTING THE LIVES AND SECURITY OF AMERICAN CITIZENS ABROAD

6-15. Rangers may be used to conduct rapid rescue and recovery of designated personnel or material over extended distances. For example—

- In April 1980, elements of the 1st Ranger Battalion participated in Operation EAGLE CLAW to rescue and evacuate American hostages held captive in Iran.
- In October 1983, elements of the 1st and 2d Ranger Battalions conducted a strategic airborne assault on Grenada (Operation URGENT FURY). Part of their mission was to rescue and evacuate American students at the True Blue medical campus.
- In December 1989, the Ranger Regiment conducted simultaneous airborne assaults into Panama during Operation JUST CAUSE. Part of its purpose was to capture General Manuel Noriega and

set the conditions to install the duly elected legitimate government. As a result, this action provided increased protection for American citizens who had been intimidated by General Noriega's forces.

6-16. Rangers are prepared to conduct regiment-level NEOs in conjunction with other conventional and special operations forces. During NEOs, the regiment battle staff, operating as a SOTF, can work directly with highly specialized joint communications, medical, air, and naval units to control each phase of the operation.

PUNITIVE ATTACKS

6-17. When a ground force is required, the Ranger Regiment can be used to conduct force projection raids against—

- Strategic or operational targets that are of high value to nation-states.
- Paramilitary organizations that sponsor terrorism.
- Illegal narcotics trade or proliferation of weapons of mass destruction.

TYPICAL RANGER TARGETS

6-18. Rangers normally are employed against targets and under conditions that require unique skills and capabilities. Ranger forces normally are assigned missions that cannot be performed by conventional infantry units. Although Rangers normally are oriented on offensive operations, they also may be employed to seize and then control key or decisive terrain. The Ranger force may conduct defensive operations associated with protection of a forward operating base or SOTF.

Haditha Dam

On 1 April 2003, elements of the 3d Ranger Battalion stormed the Haditha Dam complex northwest of Baghdad. Haditha Dam was a key objective early in Operation IRAQI FREEDOM. The 3d Ranger Battalion was tasked with preventing its destruction by hostile forces, an act that would have resulted in a humanitarian and environmental disaster and a strategic delay for coalition forces. The Ranger force seized the dam, which measured 8 kilometers across at its widest point, and held it against enemy counterattacks and artillery and mortar barrage until 6 April 2003, when control of the surrounding area was established. The dam's capture and subsequent control by Rangers denied the enemy a key crossing point over the Euphrates River, and ensured its continued use by coalition forces.

6-19. A typical Ranger battalion or regiment mission involves seizing an airfield for use by follow-on conventional forces or other SOF, and conducting raids on key targets of operational or strategic significance. Ranger operations typically differ from conventional operations in degree of risk, infiltration means (independent from friendly support), and access to operational intelligence.

6-20. Normally, Rangers are assigned targets through the controlling headquarters targeting process, which is usually the joint target board at JTF level or the targeting process at JSOTF level. Targets may be executed in close cooperation with conventional forces and may even be handed off to those units when the operating pace prohibits the Ranger force from executing them.

6-21. Ranger targets normally exceed the capability of other means of seizure or destruction when the mission involves recovery of personnel, material (such as evidence), or equipment from the target area, or when the mission escalates an ongoing U.S. military demonstration of national resolve. In some situations, Rangers may complement air power or be used in lieu of air power. Using Rangers gives the joint force commander a means of demonstrating national resolve by introducing forces on the ground in the area of conflict.

6-22. Typical Ranger targets include—

- Operational or strategic command and control systems and intelligence centers.
- Insurgent or terrorist command and control nodes.
- Hardened sites that render effective air attacks unfeasible.
- Key logistical centers, warehouses, ammunition complexes, or fuel pumping centers supplying logistical support to a theater.
- Integrated air defense systems.
- Key power generating and transmitting facilities and grids, hydroelectric dams, and irrigation systems.
- Key ports or rail complexes.
- Key installations or facilities, such as airfields, buildings, bridges, tunnels, or dams.

Chapter 7

Fires

This chapter discusses organic and nonorganic combat support functions that the Ranger Regiment typically employs to support operations. These functions include fire support, aviation, MISO, and CA.

FIRE SUPPORT ELEMENTS

7-1. Lethal and nonlethal fire support is planned, coordinated, integrated, synchronized, and supervised by the regiment's fire support element. The fire support officer is chief of the fire support element. The fire support officer and fire support element are subordinate to the regiment S-3.

7-2. The regiment fire support element contains a MAREXO, an S-6, a USAF tactical air control party, a senior fire support NCO, a current operations fire support officer, and a future operations fire support officer. The current operations fire support officer maintains situational awareness on current operations while performing the duties of assistant regimental fire support officer in garrison. The future operations fire support officer works in coordination with the senior plans officer on operations planning and coordination. Battalion fire support elements are similarly organized, except they do not have a MAREXO or a MIS officer.

7-3. Ranger Regiment and battalion fire support elements—

- Plan, coordinate, synchronize, and integrate all supporting fire support, electronic warfare, and air defense assets from all services and oversee the execution of the fire support plan. Doing so includes integrating the theater air defense early warning system into the regiment fire support plan.
- Provide staff supervision over attached or supporting fire support or air defense personnel, to include JTACs and LNOs from supporting fire support or air defense units.
- Act as integral parts of the higher headquarters targeting process to ensure Ranger fire support is synchronized, coordinated, and integrated with the JSOTF and JTF fire support plans or the higher headquarters fire support plan.
- Assist planning and executing terminal guidance operations.
- Provide a fire support cell to each of the regiment's deployed tactical command post teams while maintaining a functioning fire support element in the regiment's main command post.
- Maintain effective liaison with higher and subordinate fire support elements, as required.
- When directed, provide a joint special operations liaison element to the joint force air component.
- Facilitate the execution of fire support rehearsals at the joint, regimental, and battalion level.

FIRE SUPPORT ASSETS

7-4. The only organic indirect fire support to the regiment is a mortar platoon, of which each battalion has one (Figure 7-1, page 7-2). The platoon may employ combinations of 60-mm, 81-mm, or 120-mm mortar systems based on METT-TC. Each platoon fire direction center may split in order to support separate area of operations or missions simultaneously. Rapid and accurate mortar fire capability is enhanced by the mortar ballistic computer and guided munitions.

7-5. Rangers must rely on outside units to provide additional fire support. Rangers train with and employ Army and Air Force fire support assets consisting of rotary-wing gunships and AC-130 Spectre gunships. Rotary-wing gunships include the AH-6 from the 160th SOAR and the MH-60 defense armed penetrator.

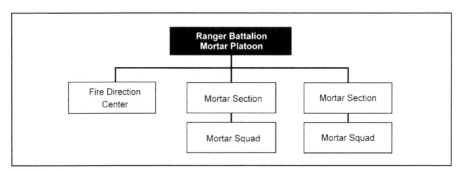

Figure 7-1. Ranger battalion mortar platoon

7-6.   The Ranger Regiment must be able to conduct terminal attacks using fixed-wing assets at the rifle platoon, reconnaissance platoon, and regimental reconnaissance team levels. In addition to special operations fire support assets, the Ranger Regiment can employ almost any means of conventional fire support assets, including naval gunfire from surface vessels; Navy, Marine, and Air Force fixed-wing close air support; artillery; and laser-guided munitions. Rangers use the latest technology to provide accurate methods of target location for precision weapons.

Army Special Operations Forces and Armor in Operation IRAQI FREEDOM

It had been more than 60 years since ARSOF and armor forces worked together in the Italian campaign of World War II. In April 2003, history repeated itself: C Company, 2d Battalion, 70th Armor Regiment, was tasked under operational control to 1/75th Ranger Battalion to help consolidate the gains of Task Force Viking (10th Special Forces Group) north of Baghdad, interdict high-value targets attempting to flee the city, and provide heavy firepower for other Ranger operations.

Given the lack of recent operational experience between two such disparate elements, the armor company commander's first task was to brief the Ranger planners on the capabilities and logistics requirements of the M1A1 Abrams tank. Of primary concern was the availability of fuel—one thirsty tank would use nearly as much fuel as an entire Ranger company's fleet of ground mobility vehicles (GMVs).

In the first Ranger/armor combined action of Operation IRAQI FREEDOM, the mission was to secure the K2 Airfield. Moving to the objective under conditions of zero illumination and near-zero visibility, the armor commander's tank rolled into a 40-foot-deep hole and overturned. After extracting the crew from the tank and caring for the wounded, the commander destroyed the tank in place. The commander then transferred to another tank to successfully continue the attack.

In a later action, Team Tank (as they came to be designated) and the Rangers assaulted the Al Sahra Airfield and Iraqi Air Force Academy to seize that terrain and interdict Highway 1 north from Baghdad. In that action the armor forces closely

---

> supported the dismounted Rangers to breach walls, suppress heavily fortified enemy positions, and destroy enemy vehicles as they appeared. In the few short weeks they worked together the Rangers and armor integrated their SOF and conventional capabilities very well.
>
> In these actions the Rangers learned to account for the difficulties of long-range travel for the tanks and to employ the tanks' ability to burst past the Ranger GMVs into the lead to engage and destroy enemy forces as they were encountered. The operations also revealed the need to reconcile the Rangers' use of infrared sights with the tankers' use of thermal sights. These successful "joint" operations validated the ability of ARSOF Soldiers and conventional forces to bridge the gap and capitalize on the strengths of each.
>
> Journal of Special Operations History
> Winter 2005

AIR DEFENSE

7-7. SOF operations typically try to exploit the advantage of air superiority over their intended target. However, if there is an enemy air threat, the regiment may use organic Stinger teams to provide local air defense in the target area. The Stinger teams are infantry Rangers cross-trained on the Stinger missile system. Each battalion maintains six current qualified two-man Stinger teams.

SPECIAL OPERATIONS AVIATION

7-8. Ranger forces rely on special operations aviation (SOA) aircraft from Army and Air Force special operations units to conduct infiltration, exfiltration, and extraction missions from target areas. SOA helicopters feature aerial-refuel capability, advanced night operability, and sophisticated navigation equipment, avionics, and electronic countermeasures. Rangers train with the same SOA units. By training with the same units, it helps to ensure the success of full-spectrum operations by the Ranger force.

7-9. Ranger forces most often use Air Force special operations fixed-wing aircraft during tactical airdrop or air-land of larger units, or to infiltrate reconnaissance teams to target areas. These aircraft are aerial-refuel capable, allowing them to operate at extended ranges. They also contain sophisticated navigation, communication, aerial delivery, and electronic countermeasure systems.

MILITARY INFORMATION SUPPORT OPERATIONS

7-10. MISO was formerly known as psychological operations or PSYOP.

7-11. Rangers develop a relationship with, and routinely employ, tactical MIS elements during full-spectrum operations.

7-12. The regiment's MIS capability comes from the Military Information Support Operations Command (MISCOM) (Airborne). A tactical MIS detachment routinely trains and operationally deploys with the regiment and with individual battalions.

7-13. The regiment receives operational control of the tactical MIS company, and each Ranger battalion typically receives a tactical MIS detachment. The detachment is made up of tactical MIS team members and a centralized HQ with an organic product support element. The disposition of MIS forces is determined by the tactical MIS OIC and NCOIC based on mission analysis and in coordination with the SR3 and J3. These tactical MIS elements help create capture/kill opportunities, advise commanders on the psychological effects of planned operations and Soldier actions, and execute military information (MILINFO) for Ranger operations. They also ensure the Ranger MISO plan is consistent with the overall MISO effort. MIS planners, when assigned for operations, are normally subordinate to the Regimental HQ and/or J3.

TACTICAL MILITARY INFORMATION SUPPORT OPERATIONS DETACHMENT

7-14. The tactical MIS detachment analyzes the higher headquarters operations order and the associated MIS tab or appendix (Appendix 3 [MISO] of Annex J [IIA] for Army operations orders/operations plans and Tab D [MIS] to Annex P [Information Operations] to Annex C [Operations] for joint operations orders/operations plans) to determine specified and implied MIS tasks. These tasks are subsequently incorporated into the supported unit MIS annex. These MIS tasks also are focused specifically on how they will support the scheme of maneuver. Therefore, the tactical MIS detachment commander normally recommends to the operations officer that he either retain his organic tactical MIS teams under tactical MIS detachment control or allocate them to subordinate units.

7-15. The tactical MIS detachment exercises staff supervision over tactical MIS teams allocated to subordinate units, monitoring their status and providing assistance in MIS planning as needed. The tactical MIS detachment is capable of providing product development, delivery, and assessment. The detachment coordinates for additional and non-organic product support when necessary. The tactical MIS teams coordinate with the detachment HQ to facilitate product development and obtain product approval.

7-16. The focus of tactical MIS detachment planning is on integrating the MIS capability and coordinating the efforts of tactical MIS elements to support the JTF commander. In the case of Ranger direct-action missions, the tactical MIS team disseminates print and loudspeaker products and develops actions for psychological effect (PSYACT) for the supported commander. Due to the limited duration of most of these direct-action missions and the time necessary to approve new products, tactical MISO product approval authority is delegated to the JTF commander. In addition, tactical MIS teams supporting direct-action missions may be allocated, from the tactical MIS detachment, a print and media production capability to provide responsive MISO products in support of time-sensitive targeting. The supported commander has the authority to broadcast command information to specific units or individuals in accordance with approved command message themes (for example, surrender instructions or noninterference instructions).

TACTICAL MILITARY INFORMATION SUPPORT OPERATIONS TEAM

7-17. In high-intensity conflict, the tactical MIS team normally provides MIS capability to a battalion. Higher rates of movement during combat operations allow tactical commanders to reinforce units in contact with MIS assets as needed. During more static or urban stability operations, planning and execution of operations are primarily conducted at the company or platoon level, which is the element that most often directly engages the local government, populace, and adversary groups, and in this instance may be assigned a tactical MIS team.

7-18. The company may require a more dedicated MIS capability to manage the population found in a company sector, particularly in urban environments when population densities are much higher (for example, 50,000 to 200,000 per company sector). Long-duration missions operating in the platoon or company area of operations allow the tactical MIS teams to develop rapport with the target audiences. This rapport can be critical to the accomplishment of their mission. The tactical MIS team leader is the MIS planner for the supported commander. He also coordinates with the tactical MIS detachment to meet the supported commander's requirements.

CIVIL AFFAIRS OPERATIONS

7-19. Rangers may employ CA Soldiers to provide CAO staff augmentation and CA planning and assessment support and assist the S-2 and other staff members with civil considerations analysis supporting preparation of the operational environment as required. This support comes in the form of a CA team (Figure 7-2, page 7-5). The team assists the Ranger force by engaging the civil component of the operational environment to minimize interference between civil and military operations, and supporting CMO to enhance mission effectiveness. The CA team conducts liaisons with civilian authorities and key leader engagement. The CA team also may coordinate with the local civilian agencies for logistical support and operational area security or base defense. Additionally, the team also may identify the locations of key electrical, hydrological, communications, or government facilities, as well as determine the effect of civilians in the operational area. The Ranger civil affairs operations staff officer (S-9) normally provides staff supervision over the attached CA elements.

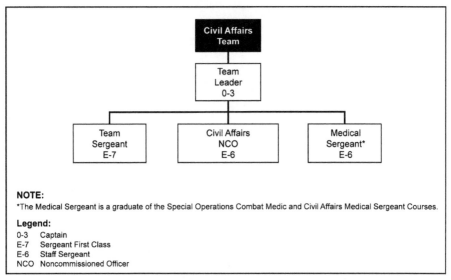

Figure 7-2. Civil Affairs team structure

This page intentionally left blank.

Chapter 8

Protection

In today's global environment, Ranger forces that operate as part of a JSOTF are assigned missions in all environments. Protection of the fighting force continues to challenge commanders. Rangers must develop 360-degree security and provide layered, overlapping, and networked capabilities that adequately protect personnel, assets, and information. Because Ranger operations depend on successful onward movement, protected personnel, and accompanying materiel; protection must occur from home station, to arrival in theater, to completion of the mission. Ranger commanders establish protection measures and identify the forces required to conduct those tasks based upon the threat assessment.

OVERVIEW OF SECURITY PLANNING

8-1. During operations, weather and terrain may be extreme and vary widely in character. The spread of urban environments and mix of civilians, paramilitaries, insurgents, and others in close physical proximity will challenge all aspects of sustainment operations. The logistics units assigned to the Ranger Regiment must be equipped to provide their own Levels I and II defense.

8-2. Tactical logistics organizations are normally the units least capable of self-defense against an opposing combat force. Given the contemporary operational environment, often they are also the targets of enemy action. However, Ranger support company capabilities for self-defense oftentimes are greater than many other units when they incorporate their .50 caliber weapons and MK-19s. Ranger commanders need to realize that when time and effort are used to defend logistics units, it degrades the ability to perform their primary support mission. There needs to be a dialogue between the battalion commander and the Ranger support company commander regarding the FSB's ability to conduct sustainment operations and its protection requirements. There is a continuum of balancing requirements. As the risk of an enemy threat increases, the ability to conduct sustainment operations decreases. The battalion commander and the Ranger support company commander must determine what amount of risk to accept and then plan accordingly with as much risk mitigation as possible.

8-3. Threats to logistics units in the sustainment area or in a noncontiguous area of operation are categorized by the three levels of defense required to counter them. Any or all levels may exist simultaneously. Emphasis on specific unit defense and security measures may depend on the anticipated threat level.

8-4. A Level I threat is a small enemy force that can be defeated by the Ranger support company's perimeter defenses established within the FSB or the Ranger support company's perimeter defenses within the Ranger unit support area. A Level I threat for a typical FSB includes a squad-sized unit or smaller groups of enemy soldiers, agents, or terrorists. Typical objectives for a Level I threat include obtaining supplies from friendly supply stocks; disrupting friendly mission command and control, sustainment operations, and facilities; and interdicting friendly LOCs.

8-5. A Level II threat is enemy activities that can be defeated by the FSB or Ranger Support Company's support area, when augmented by a response force. A typical response force is a military police or Ranger platoon. Level II threats consist of enemy special operations teams, long-range reconnaissance units, mounted or dismounted combat reconnaissance teams, and partially attritted small combat units. Typical objectives for a Level II threat include the destruction, as well as the disruption, of friendly mission command and control, logistics and commercial facilities, and the interdiction of friendly LOCs.

8-6. A Level III threat is beyond the defensive capability of the SOTF and any local reserve or response force. It normally consists of a mobile enemy force. The friendly response to a Level III threat is a Ranger combat force. Possible objectives for a Level III threat against the Ranger logistics units include destroying friendly logistics assets, supply points, command posts, arming and refueling points, and interdicting LOCs and major supply routes.

FORCE HEALTH PROTECTION SUPPORT

8-7. The Ranger Regiment has a medical section for which the regiment's surgeon has supervisory oversight. The surgeon's oversight includes responsibility for all FHP training and opportunities in the regiment. The regiment's medical section provides FHP support for the regiment's HHC, Regimental Reconnaissance Company, and regimental training detachment. It also plans and coordinates theater FHP support for the ISB and FSB operations and medical support at the Ranger objective. This support encompasses FHP augmentation; ground, rotary- and fixed-wing aircraft evacuation; and Class VIII resupply. Additionally, each Ranger battalion has a medical section with a surgeon who has supervisory oversight. Both regiment and battalion medical staffs have experience planning and leading joint casualty collection points that are routinely used during airfield seizures.

8-8. The Ranger FHP mission is to provide combat trauma management to treat the wounded and to save lives; to plan and conduct casualty evacuation for Ranger operations; to conduct a daily sick call; to plan, conduct, and instruct FHP training for individual Rangers and medical personnel; and to manage FHP administrative duties for all assigned personnel.

8-9. In addition to meeting all training standards for assignment to the Ranger Regiment and conducting individual and collective training with their assigned Ranger unit, Ranger medics are highly trained specialists and attend and conduct a significant amount of specialized trauma-management training.

TACTICAL MEDICAL EVACUATION

8-10. Ranger forces have limited casualty evacuation assets and must rely on the 528th Sustainment Brigade (Special Operations) (Airborne), theater sustainment command, and Army Service Component Command aviation support for air medical evacuation. The only tactical evacuation means Ranger forces have in the target area are a limited number of medical SOF vehicles. These vehicles are capable of carrying six litter patients each. Ranger medics have a habitual training relationship with other SOF units that have some medical ground evacuation platforms that augment the Ranger capability on a regular basis.

8-11. Generally, wounded Rangers are moved in the local target area by buddy-carry or by medical special operations vehicles to a casualty collection point where triage and trauma management occurs. The casualty collection point is normally located near a rotary-wing aircraft landing zone or fixed-wing aircraft parking area on a target airfield. The wounded are loaded onto air assets at or near the target for evacuation to the ISB or other trauma facility.

RELIGIOUS SUPPORT

8-12. The Ranger Regimental headquarters has a unit ministry team for which the regimental chaplain provides supervisory oversight. The regimental unit ministry team provides religious support and pastoral care for the regimental headquarters, the Regimental Reconnaissance Company, and the regimental training detachment. Each Ranger battalion also possesses—as an organic asset—a battalion unit ministry team. The regimental chaplain provides direction and supervision for the collective Ranger Regimental unit ministry team.

8-13. Ranger unit ministry teams provide worship services, religious rites, sacraments, ordinances, pastoral care, counseling, and crisis and emergency ministry to assigned Rangers and their family members, as well as to authorized civilian employees. To extend their ministries during decentralized operations, Ranger unit ministry teams may develop a support network of Rangers who serve as unit religious coordinators at company and platoon levels.

Chapter 9

Sustainment

To enable self-sustainment, the Ranger Regiment headquarters has a Ranger Support Operations Detachment (RSOD), and the battalions have organic Ranger support companies to enhance the expeditionary capabilities of the regiment. The regimental headquarters receives logistical support from the Ranger Special Troops Battalion in garrison or from the closest Ranger support company when deployed. With this structure, Rangers are organized with the self-sustainment capability to support internal needs for fuel, ammunition, FHP, maintenance, water production, aerial delivery, and common supplies. This capability reduces their reliance on higher logistics organizations for anything other than replenishment operations or to provide a capability not resident within the Ranger support companies. This chapter addresses the tactical employment of the Ranger Regiment logistics units and their organizational duties. It also provides an understanding of Ranger sustainment operations.

REGIMENTAL S-4

9-1. The regimental S-4 coordinates daily logistics requirements, planning, and coordination for all external support requirements. It provides operational guidance to the regiment, and maintains interface with the CONUS-based and theater management functions. The regimental S-4 coordinates with USASOC, 528th SB (SO) (A), ALE, and JTF headquarters to ensure support and sustainment requirements are properly designated.

9-2. The property book officer provides asset visibility for the entire regiment and property book accountability for the regimental headquarters and all the companies of the Ranger Special Troops Battalion.

SUPPORT OPERATIONS DETACHMENT

9-3. The RSOD, under the direction of the support operations officer, provides centralized, integrated, and automated command and control for all logistics operations within the regiment. The RSOD provides information, input, or feedback to the regimental S-1 and S-4 to plan, coordinate, and provide the battalion commander a logistical common operational picture (LCOP).

9-4. The RSOD can synchronize and provide logistics oversight to all Ranger support companies for supporting and sustaining the regiment, as well as other SOF, as directed. It possesses the capability and expertise to integrate the regiment's logistics assets into the Army Service Component Command logistics structure while simultaneously supporting the regimental headquarters. The following paragraphs explain the RSOD's mission, capabilities, and several functional elements within the logistics arena.

MISSION

9-5. The RSOD coordinates with logistics operators and FHP personnel in the fields of supply, maintenance, and movement management for the support of all units assigned or attached. Its primary concern is customer support and increasing the responsiveness of support provided by subordinate units. The detachment continually monitors the support and advises the commander on the ability to support future tactical operations. With emerging technologies such as Global Combat Support System-Army (GCSS-A), battle command sustainment support system (BCS3), Blue Force Tracker, and movement tracking system, the support operations section has access to more information and receives near-real-time information.

Therefore, support operations possess the capability to view the tactical and LCOP in the regiment. This allows support operations to identify problems more quickly and allocate resources more efficiently. BCS3 gives support operations the visibility of the logistics status from the Ranger battalion to the regiment, and potentially throughout the world depending upon the level of detail required. The RSOD serves as the first point of contact for supported units' logistics requirements. The RSOD—

- Conducts continuous regiment-focused logistics preparation of the theater.
- Plans and coordinates for aerial resupply and develops logistics synchronization matrixes.
- Submits logistics forecasts to the supporting organization.
- Coordinates and provides technical supervision for the Ranger support company's sustainment mission, which includes supply, maintenance, and coordination of transportation assets.
- Identifies tentative force structure and size to be supported.
- Coordinates the preparation of the support operations estimate on external support.
- Provides support posture and planning recommendations to the regimental commander.
- During regimental headquarters deployment, sets up and supervises the logistics center located in the tactical operations center.
- Coordinates with the S-3 to determine air routes for supply and aeromedical evacuation support.
- Provides centralized coordination for units providing support to the regiment.
- Analyzes the impact of BCS3 reports.
- Analyzes contingency mission support requirements.
- Revises customer lists (as required by changing requirements, workloads, and priorities) for support of tactical operations.
- Coordinates external logistics provided by subordinate units.
- Advises the regimental commander on the feasibility of support missions and of shortfalls that may impact on mission accomplishment.
- Serves as the single point of coordination for supported units to resolve logistics support problems.
- Plans and coordinates contingency support.
- Develops supply, service, maintenance, and transportation policies that include logistics synchronization and maintenance meetings.
- Plans and supports replenishment operations for all regimental units.

9-6. The support operations sergeant—

- Conducts continuous logistics preparation of the theater.
- Analyzes trends and forecasting requirements for supplies and equipment based on priorities and procedures.
- Coordinates major end item resupply activities within the regiment.
- Coordinates activities internal to the support operations section.

CAPABILITIES

9-7. The RSOD is assigned to the 75th Ranger Regimental headquarters. The RSOD plans, coordinates, synchronizes, and integrates logistics for the regiment and subordinate battalions, including providing subject-matter expertise in quartermaster, transportation, and ordnance operations. It also provides contingency contracting support at headquarters level to deployed Ranger elements.

9-8. The RSOD facilitates support and sustainment planning for Ranger logistics operations, as required. The RSOD provides liaison and planning elements to ensure connectivity with theater, host nation, joint, and coalition logistical infrastructures. Liaison capabilities include identifying Ranger logistics and FHP requirements, conducting logistics support planning, coordinating for resources to satisfy requirements, and arranging access to CONUS wholesale points.

SUPPLY AND SERVICES CELL

9-9. The supply and services cell plans and recommends the allocation of resources in coordination with the supported chain of command. It forecasts and monitors the distribution of supplies within the regiment. This information is entered into BCS3 at the S-4 and transferred to BCS3 at the support operations detachment. Doing so allows the support operations leader to identify problems quickly and allocate resources more efficiently through BCS3. The maintenance section provides field maintenance to the regimental headquarters. The supply and services cell also—

- Conducts continuous logistics preparation of the theater.
- Determines petroleum and water requirements.
- Provides technical expertise on supply and distribution of petroleum and water.
- Reviews bulk-fuel forecasts and adjust the forecasts after coordination with the S-3 on the impact of tactical operations on fuel requirements.
- Secures additional fuel and water storage capacity.
- Monitors water source requirements.
- Provides technical guidance on water treatment, storage, distribution, and quality-control operations.
- Provides technical expertise on supply and field-service support.
- Coordinates field-service support.
- Coordinates with Army Service Component Command–designated sustainment brigade relative to requirements for evacuation of remains to CONUS.
- Coordinates and monitors all transportation within the regimental operational environment.
- Conducts battle staff inspections to resolve problem areas and provides supply functional expertise.
- Provides advice on the management of the authorized stockage list.
- Provides technical guidance on stock records, materiel control, and accounting functions.
- Uses summary management reports to evaluate the efficiency of supply functions.
- Analyzes data and reports to determine efficiency of operations conformance to standards and trends.
- Determines operations materials-handling equipment requirements.
- Monitors subsistence supply, storage, and distribution operations in subordinate units.

RANGER SUPPORT COMPANY

9-10. The Ranger support companies are multifunctional logistics companies organic to each Ranger battalion within the Ranger Regiment. They provide field maintenance; Class I, II, III (P) (B), IV, V, VII, VIII, and IX supply; water production with limited distribution; transportation; aerial delivery; bare-base support; property management; limited CBRN decontamination and reconnaissance; and food service. The company commander provides command and control of the Ranger support company with logistics oversight and expertise provided by the RSOD. The Ranger support company supports operations at the ISB and provides necessary logistics assets to allow Ranger units of any size to be self-sustaining once they move from the ISB to SOTFs and MSSs.

MISSION

9-11. The Ranger support company commander is the senior logistics commander at battalion level. He assists the battalion S-1 and S-4 with the logistics planning. He also provides information and feedback to the battalion S-1/S-4 for their use in providing the battalion commander an LCOP.

9-12. The Ranger support company is the primary provider of common user logistics for all forces assigned or attached to the battalion. It coordinates logistics requirements with the RSOD and JTF headquarters. The Ranger support company is joint and multinational capable. It can accept augmentation of, and employ, common user logistics assets from other Services and nations and integrate their capabilities into a cohesive plan that supports the operational concept. The Ranger support company is capable, with replenishment, of supporting all of the battalion's logistical requirements. When component forces are assigned to a SOTF, they will deploy with their organic support packages for Service-specific requirements and logistics support.

9-13. The Ranger support company commander can execute the logistics plan in accordance with the battalion commander's guidance as developed by the battalion S-1 and S-4. The Ranger support company commander responds directly to the battalion executive officer, who serves as the battalion logistics integrator and assists the battalion S-1/S-4 in logistics synchronization and troubleshooting. His duties may require direct interface with the RSOD; SB (SO) (A) ALE; joint and multinational forces, other SOF; and the theater sustainment command.

9-14. The Ranger support company depends upon the battalion and other units for the following support:

- Resupply or augmentees to provide for a shortfall or capability that is not organic to the Ranger support company, or as required by METT-TC.
- Mortuary affairs planning, collection, processing, and evacuation.
- Shower, laundry, and clothing renovation that are not organic to the Ranger support company. Support can be provided by the 528th SB (SO) (A) or theater sustainment command.
- Appropriate elements to provide FHP, finance, religious, personnel, and administrative support.
- Explosive ordnance disposal capability.
- Aeromedical evacuation support.
- An LCOP for logistics outside the Ranger support company's area of operation.
- Intelligence matters.
- Common tactical picture and supported unit and echelon logistics picture.

CAPABILITIES

9-15. The support operations normally are performed by the executive officer or one of the other company officers using METT-TC. His duties include the following:

- Provide continuous battle-tracking.
- Ensure accurate, timely tactical reports are received by the command post.
- Assume command of the company, as required.
- Assist in preparation of the company operations order for the commander and, in coordination with the battalion S-1/S-4, assist with developing the concept of support for the battalion operations order.
- Conduct tactical and logistical coordination with higher, adjacent, and supported units, as appropriate.
- Oversee the development of the daily logistics packages.
- Ensure that troop-leading procedures are used to plan, prepare for, execute, and then assess the combat logistics patrol that takes the logistics packages out to the supported Ranger elements.
- As required, assist the commander in issuing orders to the company, headquarters, and attachments.
- Conduct additional missions, as required, which may include serving as the officer in charge for the quartering party, company movement officer, or company training officer.
- Assist the commander in preparations for follow-on missions.

9-16. The tasks that were accomplished previously by a RSOD support operations officer are still valid. However, the support operations tasks will also be required at the battalion and are now accomplished by the Ranger support company. These additional tasks are not to conflict with the battalion S-1/S-4 in their roles as the battalion's planners for logistics/human resources.

HEADQUARTERS DETACHMENT

9-17. The Ranger support company headquarters provides command and control, unit administration, internal supply support, billeting, discipline, security, training, and administration to assigned and attached personnel. It ensures that subordinate elements follow the policies and procedures prescribed by the battalion and Ranger support company commanders. It directs the operations of its subordinate sections as well as the overall logistics operations, less medical, in support of the battalion. The company can simultaneously establish and operate three FSBs with limited logistics support. When multiple locations are necessary, the Ranger support company executive officer plays an increased role to ensure the unit's ability to effectively operate from more than one Ranger support location.

COMMANDER

9-18. The Ranger support company commander reports to the battalion commander on the discipline and combat readiness of his Soldiers and the training of his company's tasks. His duties include the following:

- Provide input to the battalion logistics estimate and logistics annex in coordination with the battalion S-1/S-4.
- Keep the RSOD abreast of the battalion's logistics readiness posture and request backup support when needed in coordination with the battalion S-4.
- Recommend support priorities in coordination with the battalion S-4 and enforce priorities received from higher headquarters.
- Coordinate with the battalion S-2/S-3 on support locations in coordination with the battalion S-1/S-4.
- Plan and execute contingency operations, as required.
- Plan, coordinate, and control allocation of available resources in coordination with the battalion S-4, and as directed by the battalion commander's priorities of support.
- Coordinate and provide technical logistics supervision to the battalion and advise the battalion S-4 of any issues.
- Monitor battalion logistics situation report/logistics statistics and coordinate efforts with the battalion S-4.
- Plan future operations in coordination with the battalion S-1/S-4.
- Establish the logistics synchronization matrix in coordination with the battalion S-1/S-4.

FIRST SERGEANT

9-19. The first sergeant (1SG) is the commander's primary tactical advisor. He is the company's primary internal logistics operator and helps the commander plan, coordinate, and supervise all logistics activities that support the company's mission. 1SG duties include but are not limited to the following:

- Provide the commander information on the status of enlisted matters.
- Ensure the health, morale, and welfare of the unit.
- Serve as the company's senior enlisted master trainer. He is critical in identifying training requirements for individuals, crews, battle staff, units and leaders. He ensures training solutions are resourced, executed, and assessed to satisfy METL and battle tasks.
- Recommend enlisted assignments to the Ranger support company commander.
- Plan and supervise the company defense effort before, during, and after the operation.
- Supervise, inspect, and observe all matters designated by the commander.
- Assist in planning, rehearsing, and supervising key logistical actions in support of the tactical mission. These activities include resupply of Class I, III, and V products and materiel; maintenance and recovery; medical treatment and evacuation; and replacement/regimental training detachment processing.
- Assist and coordinate with the support operations in all critical functions.
- Receive incoming personnel and assign them to subordinate elements as needed.
- In conjunction with the commander, establish and maintain the foundation for company discipline.

SUPPLY SERGEANT

9-20. The supply sergeant requests, receives, issues, stores, maintains, and turns in supplies and equipment for the company. He coordinates all supply requirements and actions with the 1SG and the Ranger support company command post. Normally, the supply sergeant is supervised by the company 1SG and assisted with management for daily operations by the battalion S-4 noncommissioned officer in charge (NCOIC).

CHEMICAL, BIOLOGICAL, RADIOLOGICAL, AND NUCLEAR NONCOMMISSIONED OFFICER

9-21. The CBRN NCO assists and advises the company commander in planning for and conducting operations in a CBRN environment. He plans, conducts, coordinates, and/or supervises CBRN defense training with the 1SG. He covers such areas as decontamination procedures and use and maintenance of CBRN-related equipment. The CBRN NCO also—

- Assists the commander in developing the company operational exposure guide in accordance with the operational exposure guide from higher headquarters.
- Oversees management, training, and professional development of decontamination and reconnaissance teams.
- Makes recommendations to the commander on CBRN survey and/or monitoring, decontamination, and smoke support requirements.
- Requisitions CBRN-specific equipment and supply items.
- Assists the commander in developing and implementing the company team CBRN training program.

9-22. The CBRN NCO ensures that the training program covers the following requirements:

- Effective sustainment training in CBRN common tasks is provided by first-line supervisors.
- CBRN-related leader tasks are covered in sustainment training.
- CBRN-related collective tasks are covered in overall unit training activities.
- CBRN factors are incorporated as a condition in the performance of METL tasks.
- Company elements are inspected to ensure CBRN preparedness and the findings are provided to the commander.
- Information on enemy and friendly CBRN capabilities and activities, to include attacks, is processed and disseminated.
- Commander is up-to-date on contamination avoidance measures.
- Decontamination operations are coordinated and supervised.

DECONTAMINATION AND RECONNAISSANCE TEAMS

9-23. Each Ranger support company has two decontamination and reconnaissance teams. The teams are able to support their battalions in CBRN operations at two locations simultaneously. The decontamination and reconnaissance teams also can make positive identification and perform decontamination of most known CBRN agents.

ARMORER

9-24. The armorer performs organizational maintenance on the company's small arms and evacuates weapons as necessary to the maintenance platoon for field maintenance. He also normally assists the supply sergeant in his duties. As an option, the armorer may serve as the driver of the 1SG's vehicle.

SUSTAINMENT PLATOON

9-25. The sustainment platoon provides the battalion a single source for Class I (water), II, III (bulk) (P), IV, V, VI, VII, VIII (medical), and IX supply support to the battalion operations. The sustainment platoon receives, stores (limited), and issues Class II, III (P), IV, and IX. It also receives and distributes, in coordination with the transportation platoon, Class I and VI from the field-ration issue point, and receives and issues Class VII, as required. The platoon also maintains limited Class II, III (P), IV, and IX authorized stockage list for the battalion. The ammunition transfer holding point section supports the battalion with Class V. The platoon headquarters maintains Standard Army Management Information System (STAMIS) Standard Army Retail Supply System 1 (SARSS 1) and provides food service for assigned and attached units. The supply platoon leader also—

- Provides command and control of the sustainment platoon.
- Manages property accountability for the commander for all equipment assigned to the platoon.
- Manages the sustainment operations for the battalion.
- Provides Class I, II, III (P), and IV support to battalion.
- Receives, stores, and issues Class II, III (P), and IV.
- Maintains authorized stockage list for Classes II and III (P), and receives and issues Classes VII and IX, as required.
- Provides exchange for reparable items.
- Maintains supply STAMIS (SARSS 1 or GCSS-A).
- Provides Class V distribution in coordination with the transportation platoon to battalions and companies.
- Coordinates supply and field services support with RSOD and the battalion S-4.
- Coordinates with RSOD for augmentation as required in coordination with the battalion S-1/S-4.
- Reviews and recommends authorized stockage list changes to RSOD and the regimental S-4 through the battalion S-4.
- Plans and supervises resupply operations in coordination with the battalion S-4.
- In coordination with the battalion S-4, monitors unit combat loads to anticipate replenishment actions.
- Provides supply status report collection.
- Maintains current status of critical supplies.
- Monitors the controlled supply rate and supported units' combat loads.
- Requests through the battalion S-4 and coordinates with the RSOD for field services requirements and augmentations.
- Monitors activities within battalion for compliance with the battalion service support annex.

9-26. The sustainment platoon sergeant is the platoon's second in charge and answers to the platoon leader for the leadership, discipline, training, and welfare of the platoon's Soldiers. He coordinates the platoon's maintenance and logistical requirements, and handles the personal needs of individual Soldiers. The platoon sergeant executes the support mission with the concept of support, the operations order, and the platoon leader's guidance. He emplaces the platoon defensive sector and provides training on platoon weapons, squad and platoon tactics, and convoy defense.

SUPPLY SUPPORT ACTIVITY

9-27. The supply support activity uses SARSS 1 and related automated systems to provide authorized stockage list stock control, receipt, storage, and issue functions for both Army-common and SOF-specific items in garrison and deployed locations. The stock-control supervisor must ensure that daily start-up and closeout procedures are followed in accordance with the schedule of operations established by the supporting headquarters. Automated document processing and warehousing operations are conducted in accordance with Army Regulation (AR) 710-2, *Supply Policy Below the National Level*. The supply support activity—

- Operates the SARSS 1 system.
- Maintains a current authorized stockage list of all supported commodities.
- Processes receipts and requests for issues and turn-ins.
- Provides material release instructions to the warehouse section.
- Processes turn-ins to maintenance for reparable items.
- Performs periodic location surveys to ensure location accuracy.
- Processes inventory adjustments and creates necessary reports.

- Maintains coordination and provides general supervision over supporting signal assets.
- Establishes and performs receipt, storage, and issue for all supported commodities.
- Coordinates with support operations for delivery/pickup of issued assets and turn-ins (for maintenance or disposal).
- Performs storage and inventory management activities as directed by stock control.

CLASS III SECTION

9-28. The petroleum, oils and lubricants section provides the management, stock, and delivery of all Class III (B) (P) items to the battalion.

AMMUNITION SECTION

9-29. The ammunition section manages the ammunitions and explosives training, basic load, and operational requirements. The section ensures that transportation of ammunition and explosives is achieved by standing operating procedures and Army and Air Force regulations. It is capable of operating one ammunition transfer point.

WATER SECTION

9-30. The water section provides up to 2,500 gallons per hour of potable water daily through the operation of three 125-gallon-per-hour reverse-osmosis lightweight water purifiers operating 20 hours per day. An additional capability exists to purify up to 6,000 gallons of water per day using the LS3 ultraviolet system. The section provides a limited distribution using the Forward Area Water Point Supply System. The section is capable of storing a maximum of 9,000 gallons of water.

FOOD SERVICES SECTION

9-31. Class I is provided by the food service section. This section provides food service and food preparation for the battalion and organic and attached personnel in the SOTF. The Ranger support company food service section and the food service capability of supported units merges with the battalion food service section to form a consolidated messing facility. It distributes prepackaged and prepared food. The food service section has the ability to prepare two unitized group ration (UGR) meals per day—one UGR heat-and-serve meal and one UGR A-ration meal (dependent upon METT-TC) using organic food service equipment. The food operations section has the organic capability to prepare subsistence for consolidated and remote feeding operations, manages the unit basic load of Class I, and provides food service support for 800 Soldiers and assigned personnel. The food operations section operates in both garrison and field environments.

PROPERTY BOOK TEAM

9-32. This team provides asset visibility and property book accountability for the supported Ranger battalion, and operates in garrison and deployed environments.

DISTRIBUTION PLATOON

9-33. The distribution platoon provides transportation of materiel and personnel, movement control, and aerial-delivery support functions. The distribution platoon leader—

- Provides command and control of the distribution platoon.
- Manages property accountability for the commander for all equipment assigned to the platoon.
- Manages distribution for the battalion.
- Manages ground transportation assets and operations.
- Manages aerial-delivery operations.
- Coordinates and monitors the movement of replenishment stocks/services for the Ranger support company.

- Monitors retrogrades of aerial-delivery equipment.
- Coordinates retrograde of equipment and supplies with the RSOD in coordination with the battalion S-4.
- Coordinates delivery priorities with the battalion S-4.
- Coordinates supplemental transportation in support of the battalion in coordination with the battalion S-4.
- Determines requirements and plans for air-resupply operations in coordination with the battalion S-4.

9-34. The distribution platoon sergeant is the platoon's second in charge and answers to the platoon leader for the leadership, discipline, training, and welfare of its Soldiers. He coordinates the maintenance and logistics requirements, and handles the personal needs of each Soldier. The platoon sergeant executes the support mission with the concept of support, the operations order, and platoon leader's guidance. He emplaces the platoon defensive sector and ensures training on weapons, squad and platoon tactics, and convoy defense. The platoon sergeant also oversees the following elements:

- Truck squad—
  - Provides the capability for transport of supplies and equipment.
  - Provides motor transport capable of moving containerized and noncontainerized cargo.
  - Hauls an entire Ranger company.
  - Provides movement control and aerial delivery support.
- Movement control team—
  - Provides the management and coordination for movement control.
  - Coordinates the loading, offloading, and transport of supplies, ammunition, explosives, equipment, materials-handling equipment, and oversized equipment to and from aircraft or other transport, as directed.
- Aerial-delivery section—
  - Prepares up to 10 tons of general supplies and equipment per day for aerial resupply.
  - Provides personnel parachute-packing support.

MAINTENANCE PLATOON

9-35. The maintenance platoon provides field-level maintenance on Army-common and SOF-specific automotive, ground-support, armament, construction, electronic/communications, quartermaster, and a wide variety of commercial equipment for the battalion and attached units. The platoon also maintains a Class IX shop stock and bench stock, as appropriate, and provides recovery support. The bare-base section provides utilities evaluation when establishing FSBs or remote marshalling bases in austere operational areas. It also supervises supported unit supplied labor in the establishment of deployment cells capable of supporting 800 Soldiers and supervises the operation of these cells once they are established. It also provides construction capability to fabricate training facilities/targets and limited combat-engineering support. The platoon operates in garrison and field environments.

PLATOON LEADER

9-36. The battalion maintenance warrant officer also performs the duties as automotive platoon leader. He controls and directs the mission, confirms maintenance jobs, prioritizes support requirements, and makes sure the commander's guidance is followed. He maintains the readiness of the platoon's personnel and equipment. He also oversees the health, welfare, and morale of platoon personnel. The unit commander primarily establishes the platoon leader's duties. The platoon leader—

- Trains the platoon personnel.
- Supervises recovery team operations, forward repair elements, or other on-site maintenance missions.
- Reviews and evaluates operator/crew preventive maintenance checks for equipment.
- Determines battalion equipment operators' licensing requirements.

- Participates in the analysis, planning, and supervising of all maintenance activities.
- Manages property accountability for the commander for all platoon equipment.
- Understands the battalion maintenance priorities, and ensures the maintenance platoon adheres to the established priorities and guidance.
- Serves as maintenance control officer.
- Recommends allocation of maintenance assets in coordination with the company commanders to the battalion S-4 in accordance with the commander's prioritization of support.
- Monitors maintenance operations and Class IX, line-replaceable unit and major-assembly replenishment.
- Reviews and recommends authorized stockage list changes to RSOD and regimental S-4 through the battalion S-4.
- Forecasts and monitors the workload for all equipment by type.
- Monitors maintenance shop production and job status.
- Intensively manages non-mission-capable high-priority jobs (critical combat-power-producing jobs) in coordination with the battalion S-4.
- Coordinates additional requirements through RSOD in coordination with the battalion S-4.
- Coordinates critical parts status with RSOD in coordination with the battalion S-4.
- Coordinates for personnel with special military occupational specialties to support slice units equipment in coordination with the battalion S-1/S-4 and the RSOD.

PLATOON SERGEANT

9-37. The maintenance platoon sergeant is the platoon's second in charge and answers to the platoon leader for the leadership, discipline, training, and welfare of its Soldiers. He coordinates the platoon's maintenance and logistical requirements, and handles the personal needs of individual Soldiers. The platoon sergeant executes the support mission of the platoon with the concept of support, the operations order, and the platoon leader's guidance. He emplaces the platoon defensive sector and makes sure the platoon is trained on weapons, squad and platoon tactics, and convoy defense.

AUTOMOTIVE MAINTENANCE LEADER

9-38. The battalion maintenance warrant officer provides technical expertise on all aspects of the field-maintenance mission. He uses his advanced diagnostics and troubleshooting skills to isolate system faults and expedite the repair and return of major weapons systems. Because of his technical expertise, the maintenance warrant officer advises the commander on all matters pertaining to battle damage assessment reports. The automotive maintenance commander—

- Provides input to the Ranger support company and battalion commander's plans.
- Organizes and allocates resources to execute the field-maintenance mission in support of Army-common and SOF-specific automotive, ground support equipment, armament, construction, electronic/communications, quartermaster, and a wide variety of commercial equipment for the battalion and attached units.
- Evaluates and inspects maintenance operations and develops and implements corrective action plans, where necessary, to comply with regulatory and statutory requirements applicable in garrison and field environments.
- Identifies technical training shortfalls and, when necessary, trains maintenance personnel to accurately diagnose/troubleshoot mechanical, electrical, pneumatic, and hydraulic malfunctions accurately using the latest equipment, technical publications, and procedures available.
- Provides management oversight and technical guidance on the establishment of unit stocks of combat spares in accordance with applicable supply regulations.
- Coordinates for or, as necessary, provides technical training for Standard Army Maintenance System 1 Enhanced (SAMS1-E) operators and repair parts specialists (92A).
- Assists in the development and updating of the field-maintenance standing operating procedures as they pertain to the conduct of field-level maintenance operations.

- Oversees the unit's test, measurement, and diagnostic equipment programs and ensures the programs are covered in the field-maintenance standard operating procedures and meet the regulatory guidance.
- Ensures that recovery-vehicle operators are properly trained and certified to perform recovery operations.
- Uses automated maintenance management systems to provide maintenance information to the commander and higher headquarters.
- Assists in the planning, scheduling, and publishing of the scheduled service plan for all assigned equipment per the applicable technical manual/lubrication order.
- Conducts technical inspections of unit equipment to determine the equipment maintenance status.
- Enforces the maintenance of up-to-date technical publications for use by maintenance personnel.
- Establishes the commander's quality-assurance program for maintenance and repairs, and oversees all quality-control inspections and inspectors to validate their capability to identify improper repairs and scheduled services.
- Serves as the unit's point of contact for automated readiness-reporting and mileage-reporting issues.
- Evaluates and ensures the quality of maintenance completed.
- Develops a training and cross-training plan for maintenance personnel.
- Coordinates the recovery of battalion equipment.
- Monitors the status of equipment undergoing repairs and determines the status of Class IX repair parts.
- Plans for continuity of maintenance support during periods of movement.
- Manages production control, to include the assignment of work to shop sections and the compilation of prescribed reports and records.
- Coordinates maintenance section, recovery and service section, and maintenance team requirements for using recovery section assets.
- Coordinates the activities of the inspectors and maintenance personnel to ensure maintenance standards.
- Executes maintenance priorities as established by the battalion commander.
- Anticipates expected workloads, shop progress, difficulties encountered during repair actions, and maintenance supply actions.
- Analyzes and plans all maintenance activities.
- Coordinates field-maintenance requirements with the battalion S-4 and RSOD, as appropriate.
- Develops the maintenance services plan for battalion equipment.
- Develops and executes the battalion licensing program.
- Integrates Army Service Component Command maintenance teams into the Ranger support company.

FIELD MAINTENANCE SECTION

9-39. This section provides field-level maintenance on Army-common and SOF-specific automotive, electronics and communications, ground-support, armament, construction, quartermaster, and a wide variety of commercial equipment for the battalion and attached units, and provides reinforcing maintenance to the forward maintenance team (FMT). This section also assists with organizational services on selected pieces of equipment organic to the Ranger support company and the battalion. While performing services, the mechanic completes a DA Form 5988-E (Equipment Inspection Maintenance Worksheet) with the assigned operator and his supervisor. He then submits the form to the SAMS1-E operator.

9-40. Maintenance advances, such as the multi-capable mechanic, advances in diagnostics and prognostics maintenance capabilities, and the introduction of the battalion repair system, enhance the Ranger support company maintenance platoon's capabilities.

FORWARD MAINTENANCE TEAM

9-41. The maintenance section can field one FMT that is organized to provide field maintenance for all vehicles organic to the battalion companies. The Ranger support company commander sets the FMT priorities in accordance with the battalion commander's guidance. When deployed in support of a Ranger company, the FMT operates under operational control of the company 1SG, and the maintenance NCOIC supervises the team. The scope and level of repair is based on METT-TC. The FMT makes repairs as far forward as possible, and returns the piece of equipment to the unit. During combat, the FMT performs battle damage assessment, diagnostics, and on-system replacement of line-replaceable units. Emphasis is placed on troubleshooting, diagnosing malfunctions, and fixing the equipment by component replacement. If the tactical situation permits, the FMT focuses on completing jobs on-site.

9-42. The FMT carries limited on-board combat spares to help facilitate forward repairs. If inoperable equipment is not repairable, due either to METT-TC or a lack of repair parts, the team uses recovery assets to assist the maneuver company and may recover inoperable equipment to the unit maintenance collection point or designated linkup point. The FMT is fully integrated into the Ranger company's operational plans. There are also two other elements that provide specific support that make meeting mission requirements easier:

- *Base support section*. When augmented with troop labor, each engineer section supervises and provides the engineer expertise to establish a forward support base. The section provides engineer expertise to tie-in to D-Cell or Navy construction engineer battalion support for bare-base or warm-base construction, as required.
- *CBRN technical escort detachment*. This element provides expertise for sensitive-site exploitation capability to the supported battalion. It also conducts battalion-sized decontamination operations with troop labor from the supported battalion.

RANGER LOGISTICS SUPPORT

9-43. The Ranger Regiment relies on logistic support from home station or prepackaged supplies during the initial stages of the deployment. As the theater matures, replenishment is provided by the theater sustainment command or joint logistics providers within the joint operations area. The regiment primarily receives logistics support from the RSOD and the Ranger battalion's Ranger support company.

9-44. Ranger units can deploy, fight, and sustain basic requirements for up to 12 days. With replenishment and augmentees from the theater sustainment command, the Ranger force can sustain itself indefinitely. Each Ranger battalion and the regimental headquarters company have a fighting and sustainment load of critical classes of supply and equipment located at tenant installations. This load provides the capability of 3 days' supply of ammunition and 5 days' supply of food, FHP supplies, and batteries. Also included in the load is CBRN contingency equipment and special equipment pallets to provide a limited maintenance capability. Additionally, the Ranger force maintains contingency stocks at Anniston Army Depot, Alabama that provide an additional sustainment capability of 9 days of all required supplies. The combination of the Ranger fighting, sustainment, and contingency loads provides a stockage of critical requirements providing the Ranger Regiment the capability of sustaining operations in an austere environment for 12 days.

9-45. The Ranger force is limited by airframes for transport of both personnel and equipment. Whenever possible, the contingency stocks are augmented from theater sources to reduce the number of aircraft required deploying and supporting a Ranger force.

9-46. Rangers deploy in support of an operations plan or concept plan. Therefore, the logistics concept of support must be flexible and tailored to support the operational requirement. As a member of USSOCOM, Rangers receive support from installations under Title 10, United States Code (10 USC). 10 USC requires installations where Rangers are located to provide all requirements to deploy the Ranger force.

INSTALLATION DEPLOYMENT SUPPORT ELEMENT

9-47. Installations that host Ranger units provide an installation deployment support element whose mission is to provide personnel, equipment, and material as necessary to outload and deploy Ranger units from home station. The installation deployment support element will not deploy with the Ranger unit. As a minimum, the installation deployment support element will organize logistics, transportation, operations, and training support, as well as a Departure Airfield Control Group for deploying Ranger units. The installation deployment support element is designated to provide the support and resources to push a Ranger force out on an emergency deployment. It is controlled by the installation commander and synchronized through the installation emergency operations center during emergency deployment readiness exercises and real-world deployments.

EARLY-ENTRY LOGISTICS SUPPORT

9-48. The Ranger battalions are self-sufficient for independent operations of up to 5 days. The battalions conduct field maintenance for all organic Army-common and SOF-specific vehicles, ground support, communications electronics, weapons, night-vision, optics, lasers, and automated data processing equipment. Ranger battalions deploy with up to 3 days of ammunition. They have the ability to receive, store, and conduct tactical resupply for another 3 days of ammunition. They have dedicated and qualified ammunition handlers to ensure proper handling and accountability of ammunition in accordance with regulations.

9-49. The Ranger Regiment requires external air and ground transportation for deployment and most infiltrations. Each battalion of the regiment has vehicles for crew-served weapons, medical evacuation, communication, and logistics resupply for tactical movements in and around objective areas. The regiment has prepackaged supplies that can sustain the regiment headquarters and two Ranger battalions for 15 days, or the entire regiment for 11 days. These supplies are stored at the home stations and at Army depots. This resupply system allows the regiment to deploy rapidly and be self-sustaining until the RSOD can coordinate with the JSOTF and Army special operations liaison elements to obtain support from the theater sustainment command or joint logistics providers within the joint operations area. This system also allows deploying Rangers to take what supplies they need or the airflow will allow, and enables follow-on aircraft to quickly build up required supplies.

EXTERNAL LOGISTICS REQUIREMENTS

9-50. Installations where the Ranger battalions are stationed provide base operations support, to include the deployment of those units. Due to Ranger logistics limitations during deployment of the Ranger support company, installations provide transportation and outload assistance to Ranger units during deployment from their home station.

9-51. Installations assist other supporting elements of the regiment when the regiment marshals at one location prior to deployment to an OCONUS ISB or FSB. Support provided by these installations includes the following:

- Billeting and rehearsal areas with area security for each.
- Ground transport within the ISB or FSB.
- Centralized messing.
- Maintenance for vehicles, weapons, night-vision/optics, and communications equipment above the capabilities of the Ranger support company, as required.

SUSTAINMENT AND CONTINGENCY LOADS

9-52. Installation ammunition supply points and warehouses where Rangers are stationed hold Ranger unit basic loads of ammunition, batteries, chemical protective overgarments, and meals, ready to eat (MREs). These are commonly referred to as fighting and sustainment loads because they are configured to provide a Ranger's basic fighting load plus 20 percent for sustainment. Each Ranger battalion maintains these loads. These loads enable the battalion to deploy in 18 hours with up to a 3-day supply of Class V; a 5-day supply of Classes I, III, VIII, and IX; and sufficient batteries without requiring external resupply.

ANNISTON ARMY DEPOT CONTINGENCY STOCKS

9-53. Anniston Army Depot maintains prepackaged contingency stocks of ammunition, batteries, MREs, Joint Service Lightweight Integrated Suit Technology chemical protective overgarments, and Class VIII items for the 75th Ranger Regiment. These contingency stocks are a combination of pallets and airdrop bundles, which are easily transferred to a departure airfield by ground vehicle and then transported by the approximate lift equivalent of six C-17s. Anniston Army Depot contingency stocks can resupply the entire regiment for up to 9 days without external resupply. The regimental commander is the release authority for this asset, and this release is affected through faxing a memorandum to Anniston Army Depot from an ISB, FSB, or the regimental headquarters. This unique capability provides the regimental commander flexibility above and beyond support available in-theater, as well as being able to function in austere or undeveloped theaters.

9-54. The Anniston Army Depot contingency stocks are broken down into four categories: Battalion Sets A, B, and C, and miscellaneous pallets. The miscellaneous pallets include two sets of Joint Service Lightweight Integrated Suit Technology for every Ranger in the regiment, a 9-day stock of MREs for one battalion, and the regimental headquarters battery pallet of 9 days of stock.

9-55. Battalion Sets A and B are the same. Each provides 9 days of ammunition, batteries, and Class VIII supplies for a battalion. The sets are built upon wooden pallets, shippable in standard civilian tractor-trailers, or are placed four to one 436L pallet and hauled on flatbeds. Each pallet is broken down by Department of Defense identification code and channel reassignment function 49 compatibility group. The intended use of the pallets is support in an uncertain FSB or ISB environment where they are flown in, offloaded, and distributed by the Ranger support company. The regimental headquarters set is 9 days of stock, mirroring Sets A and B.

9-56. Set C is different from Sets A and B. Set C provides 6 days' supply of ammunition, batteries, and Class VIII supplies for a battalion, but also has an additional 20 airdrop bundle-pallets that provide 3 days' supply of ammunition, FHP supply batteries, and MREs to one battalion. Set C is intended to provide a large resupply capability in a hostile environment or where there is no airfield capable of receiving aircraft.

Note: Rangers use the statement of requirements process during training exercises when coordinating logistics and to convey needs during combat or contingency operations to planning staffs throughout the Services.

RESUPPLY METHODS

9-57. Resupply operations for Ranger units are planned during the operations order production process and tailored to meet mission requirements. Numerous resupply options exist.

9-58. If the Ranger force is inserted into the objective area by airborne or airland assault, Rangers use door bundles of critical supplies and airland bulk supplies, as required. Door bundles use A-7 and A-21 containers marked to identify different loads or units. The most common door bundle loads are Ranger antiarmor weapons system ammunition, mortar ammunition, and Stinger rounds. Resupply packages are also airlanded with the force and then combat-offloaded. These loads may include water, ammunition, demolitions, FHP supplies, and batteries. Each Ranger support company provides vehicles with trailers that airland to provide a mobile resupply capability during early-entry operations, dependant on METT-TC.

9-59. If required during the course of an airborne operation, the Ranger force receives a resupply by speedball. Speedball is a resupply technique in which tailored supplies—such as bagged water, MREs, ammunition, batteries, and FHP supplies—are configured by the Ranger support company, packed in duffel bags, and then encased in corrugated cardboard honeycomb.

9-60. These resupply packages are configured to resupply a squad and are marked in accordance with the regimental standing operating procedures. Speedballs can be delivered by ground transportation or dropped to the Ranger force from a helicopter in any terrain without having to land or hover. During continuous combat operations, the Ranger force primarily uses speedball resupply. The Ranger unit has the ability to receive cached supplies for recovery by small Ranger patrol teams.

9-61. Several other resupply options are available, but are not normally required. Among them are the container delivery system; the high-speed, low-level airdrop system; the CTU-2A; and the high-altitude airdrop resupply system or emerging aerial-delivery systems.

9-62. The container delivery system consists of multiple, individually rigged containers—each with its own parachute—weighing up to 2,000 pounds each. The system provides single-pass delivery of up to 16 containers by C-17 aircraft. The container delivery system loads may be delivered into drop zones using multiple points of impact to allow for tactical separation. The Ranger force commander can direct the use of multiple points of impact, noting the advantages to be gained against the requirement for multiple passes. The Ranger Regiment stores packed container delivery system bundles at Anniston Army Depot. Each bundle contains 3 days' supply of ammunition, FHP supplies, batteries, and MREs for one battalion.

9-63. The high-speed, low-level airdrop system is a single A-21 container specially rigged to withstand the parachute opening shock when airdropped from a fixed-wing aircraft at high speed. This system can be used to deliver up to 600 pounds for each container, with a maximum of four containers for each pass, at speeds up to 250 knots.

9-64. The CTU-2A is a high-speed, aerial-delivery container that can be used to deliver supplies from high-performance aircraft flying at a minimum altitude of 300 feet above ground level and a maximum airspeed of 425 knots. The CTU-2A is carried on the bomb racks of fighter- or bomber-type aircraft. Upon release, a pilot parachute deploys the main chute, and the container descends slowly. The CTU-2A can be used to deliver up to 500 pounds of supplies such as weapons, water, food, or munitions. The container can be destroyed by burning. The main advantage of this system is that it can be delivered by high-performance aircraft deep behind enemy lines and in a dense air-defense environment. The accuracy of this system is equal to that of a conventional bomb strike.

9-65. At times it may be better to drop resupply loads to the Ranger force from a high altitude. The high-altitude airdrop resupply system permits containerized unit loads weighing from 200 to 2,000 pounds to be delivered from aircraft at speeds up to 150 knots from up to 25,000 feet above ground level. The high-altitude airdrop resupply system consists of a cargo parachute, an airdrop container, an altitude sensor, and a pilot chute. The pilot chute gives the descending bundle a speed slightly greater than an accompanying high-altitude low-opening parachutist. The system provides for the steady free-fall descent of loads from altitudes between 2,000 and 25,000 feet to an altitude at which a barometric sensor actuates deployment of the main parachute. This steadiness allows safe and accurate delivery of loads onto unprepared drop zones. The high-altitude airdrop resupply system can deliver a payload to within 260 meters from target impact point from a 10,000 foot altitude, with a proportional degree of accuracy from 25,000 feet. Ranger units can use the system for resupply of battalions and smaller units with rations, ammunitions, and FHP supplies, breaking down the containerized material into man-packed loads.

This page intentionally left blank.

Glossary

1SG	First Sergeant
AFSOC	Air Force Special Operations Command
ALE	automatic link establishment
AM	amplitude modulation
AR	Army regulation
ARNG	Army National Guard
ARNGUS	Army National Guard of the United States
ARSOF	Army special operations forces
AS-2	assistant intelligence officer
BCS3	battle command sustainment support system
CA	Civil Affairs
CBRN	chemical, biological, radiological, and nuclear
CI	counterintelligence
COMSEC	communications security
CONUS	continental United States
COP	common operational picture
FHP	force health protection
FM	field manual; frequency modulation
FMT	forward maintenance team
FSB	forward staging base
G-2	Deputy Chief of Staff for Intelligence
G-8	Combat Development Manpower and Programming section
GCSS-A	Global Command Support System-Army
GIG	global information grid
GMV	ground mobility vehicle
HCT	HUMINT collection team
HF	high frequency
HHC	headquarters and headquarters company
HSOC	home station operations center
HUMINT	human intelligence
INMARSAT	international maritime satellite
ISB	intermediate staging base
J-2	intelligence directorate
JP	joint publication
JSOTF	joint special operations task force
JTF	joint task force
LCOP	logistical common operational picture
LNO	liaison officer

LOC	line of communications
MAREXO	Marine Exchange Officer
METL	mission-essential task list
METT-TC	mission, enemy, terrain and weather, troops and support available, time available, civil considerations
MIS	Military Information Support
MISO	Military Information Support Operations
MISCOM	Military Information Support Operations Command
mm	millimeter
MSS	mission support site
NCO	noncommissioned officer
NCOIC	noncommissioned officer in charge
NEO	noncombatant evacuation operation
OB	order of battle
OCONUS	outside the continental United States
OIC	officer in charge
RSOD	Ranger Support Operations Detachment
S-1	personnel staff officer
S-2	intelligence staff officer
S-2X	CI/HUMINT intelligence staff officer
S-3	operations and training staff officer
S-4	logistics staff officer
S-5	plans staff officer
S-6	signal staff officer
S-8	strategic plans and requirements
S-9	civil-military operations staff officer
SAMS1-E	Standard Army Maintenance System 1 Enhanced
SARSS 1	Standard Army Retail Supply System 1
SATCOM	satellite communications
SCAMPI	Single Channel Antijam Manpack Portable Interface
SEAL	sea air land
SF	Special Forces
SHF	super-high frequency
SIGINT	signals intelligence
SOA	special operations aviation
SOAR	Special Operations Aviation Regiment
SOF	special operations forces
SOTF	special operations task force
SOWT	special operations weather team
STAMIS	Standard Army Management Information System
TACSAT	tactical satellite

TCAE	technical control and analysis element
TFCICA	task force counterintelligence collection authority
TIP	target intelligence package
UAS	unmanned aircraft system
UGR	unitized group ration
UHF	ultrahigh frequency
UN	United Nations
U.S.	United States
USAJFKSWCS	United States Army John F. Kennedy Special Warfare Center and School
USAF	United States Air Force
USAR	United States Army Reserve
USASOC	United States Army Special Operations Command
USSOCOM	United States Special Operations Command
VHF	very high frequency

SECTION II – TERMS

Army special operations forces

Those Active and Reserve Component Army forces designated by the Secretary of Defense that are specifically organized, trained, and equipped to conduct and support special operations. Also called **ARSOF**.

counterterrorism

Actions taken directly against terrorist networks and indirectly to influence and render global and regional environments inhospitable to terrorist networks. Also called **CT**. (JP 3-26)

direct action

Short-duration strikes and other small-scale offensive actions conducted as a special operation in hostile, denied, or politically sensitive environments and which employ specialized military capabilities to seize, destroy, capture, exploit, recover, or damage designated targets. Also called **DA**. (JP 3-05)

joint special operations task force

A joint task force composed of special operations units from more than one Service, formed to carry out a specific special operation or prosecute special operations in support of a theater campaign or other operations. Also called **JSOTF**. (JP 3-05)

noncombatant evacuation operations

Operations directed by the Department of State or other appropriate authority, in conjunction with the Department of Defense, whereby noncombatants are evacuated from foreign countries when their lives are endangered by war, civil unrest, or natural disaster to safe havens as designated by the Department of State. Also called **NEOs**. (JP 3-68)

raid

An operation to temporarily seize an area in order to secure information, confuse an adversary, capture personnel or equipment, or to destroy a capability. It ends with a planned withdrawal upon completion of the assigned mission. (JP 3-0)

Rangers

Rapidly deployable airborne light infantry organized and trained to conduct highly complex joint direct action operations in coordination with or in support of other special operations units of all Services. (JP 3-05)

Special Forces

U.S. Army forces organized, trained, and equipped to conduct special operations with an emphasis on unconventional warfare capabilities. Also called **SF**. (JP 3-05)

special operations

Operations requiring unique modes of employment, tactical techniques, equipment and training often conducted in hostile, denied, or politically sensitive environments and characterized by one or more of the following: time sensitive, clandestine, low visibility, conducted with and/or through indigenous forces, requiring regional expertise, and/or a high degree of risk. Also called **SO**. (JP 3-05)

special operations task force

A temporary or semipermanent grouping of ARSOF units under one commander and formed to carry out a specific operation or a continuing mission. Also called **SOTF**.

special operations weather team

A task organized team of AFSOC personnel specially organized, trained, and equipped to integrate full-spectrum SOF-specific planning and execution METOC operations and support into ARSOF operational requirements. Also called **SOWT**.

special reconnaissance

Reconnaissance and surveillance actions conducted as a special operation in hostile, denied, or politically sensitive environments to collect or verify information of strategic or operational significance, employing military capabilities not normally found in conventional forces. Also called **SR**. (JP 3-05)

weapons of mass destruction

Chemical, biological, radiological, or nuclear weapons capable of a high order of destruction or causing mass casualties and exclude the means of transporting or propelling the weapon where such means is a separable and divisible part from the weapon. Also called **WMD**. (JP 3-40)

References

SOURCES USED
These are the sources quoted or paraphrased in this publication.

Army Forms
DA Form 5988-E, *Equipment Inspection Maintenance Worksheet.*

Army Publications
ADP 6-0, *Mission Command*, 17 May 2012.

AR 710-2, *Supply Policy Below the National Level*, 28 March 2008.

FM 3-05, *Army Special Operations Forces*, 1 December 2010.

Joint Publications
JP 3-0, *Joint Operations,* 11 August 2011.

JP 3-05, *Special Operations*, 18 April 2011.

JP 3-26, *Counterterrorism*, 13 November 2009.

JP 3-40, *Combating Weapons of Mass Destruction*, 10 June 2009.

JP 3-68, *Noncombatant Evacuation Operations*, 23 December 2010.

DOCUMENTS NEEDED
None.

READINGS RECOMMENDED
None.

This page intentionally left blank.

Index

This page intentionally left blank.

FM 3-75 **(FM 3-05.50)**
23 May 2012

By Order of the Secretary of the Army:

RAYMOND T. ODIERNO
General, United States Army
Chief of Staff

Official:

Joyce E. Morrow

JOYCE E. MORROW
Administrative Assistant to the
Secretary of the Army
1135411

DISTRIBUTION:

Active Army, Army National Guard, and United States Army Reserve: To be distributed
in accordance with initial distribution number 115985, requirements for FM 3-75.

This page intentionally left blank.